Real Estate Analysis in the Information Age

The creation, accumulation, and use of copious amounts of data are driving rapid change across a wide variety of industries and academic disciplines. This "big data" phenomenon is the result of recent developments in computational technology and improved data gathering techniques that have led to substantial innovation in the collection, storage, management, and analysis of data.

Real Estate Analysis in the Information Age: Techniques for Big Data and Statistical Modeling focuses on the real estate discipline, guiding researchers and practitioners alike on the use of data-centric methods and analysis from applied and theoretical perspectives. In it, the authors detail the integration of big data into conventional real estate research and analysis. The book is process-oriented, not only describing big data and associated methods, but also showing the reader how to use these methods through case studies. The running theme is the construction of efficient, transparent, and reproducible research through the systematic organization and application of data, both traditional and "big". The final chapters investigate legal issues, particularly related to those data that are publicly available, and conclude by speculating on the future of big data in real estate.

Kimberly Winson-Geideman is Senior Lecturer in Property at the University of Melbourne, Australia.

Andy Krause is Principal Data Scientist at Greenfield Advisors, USA.

Clifford A. Lipscomb is the Vice Chairman and Co-Managing Director at Greenfield Advisors, USA.

Nicholas Evangelopoulos is Professor of Business Analytics at the University of North Texas, USA.

Real Estate Analysis in the Information Age

Techniques for Big Data and Statistical Modeling

Kimberly Winson-Geideman, Andy Krause, Clifford A. Lipscomb, and Nicholas Evangelopoulos

Routledge
Taylor & Francis Group
LONDON AND NEW YORK

First published 2018
by Routledge
2 Park Square, Milton Park, Abingdon, Oxon OX14 4RN

and by Routledge
711 Third Avenue, New York, NY 10017

Routledge is an imprint of the Taylor & Francis Group, an informa business

British Library Cataloguing-in-Publication Data
A catalogue record for this book is available from the British Library

Library of Congress Cataloging-in-Publication Data
Names: Winson-Geideman, Kimberly, author.
Title: Real estate analysis in the information age : techniques for big data and statistical modeling / Kimberly Winson-Geideman [and three others].
Description: Abingdon, Oxon ; New York, NY : Routledge, 2018. | Includes bibliographical references and index.
Identifiers: LCCN 2017027241 | ISBN 9781138232907 (pbk. : alk. paper) | ISBN 9781138232891 (hardback : alk. paper) | ISBN 9781315311135 (ebook : alk. paper)
Subjects: LCSH: Real property—Data processing. | Real estate business—Data processing. | Real estate business—Statistical methods.
Classification: LCC HD1380 .W56 2018 | DDC 333.330285—dc23
LC record available at https://lccn.loc.gov/2017027241

ISBN: 978-1-138-23289-1 (hbk)
ISBN: 978-1-138-23290-7 (pbk)
ISBN: 978-1-315-31113-5 (ebk)

Typeset in Bembo
by Apex CoVantage, LLC

This book is dedicated to our respective spouses and children who, incidentally, now believe 'adulting' means writing a book.

Contents

Foreword

In 1967, John Ridker and Richard Henning published a paper in *Review of Economics and Statistics* on the effect of air pollution on home prices. This was a pioneering and influential paper – one of the first to use hedonic price analysis of property markets to attempt a cost-benefit analysis of a public project. As of this writing it has been cited 981 times. What strikes us now about this paper is that it used data on a mere 167 home sales to construct a regression model of home values on a number of structural and neighborhood characteristics including measures of air quality, to determine that there was a statistically significant relationship between the two variables of interest.

It would be a gross mischaracterization of statistical practice in 1967 to say that a sample of this size was regarded as wholly adequate. Certainly there were some near-contemporary works in real estate analysis such as Hugh Nourse's 1962 work on public housing, and John Kain and John Quigley's 1970 construction of hedonic models, whose sample sizes numbered in the thousands. Nevertheless, the Ridker and Henning's work was published, and the statistical framework and hypothesis testing procedures thought to be adequately justified.

After all (so the thinking sometimes went), 167 was, in fact, a big sample. "Big" meant mostly that you could use the Z approximation to the t-table, and statistics textbooks provided exact critical values of the t-distribution for sample sizes (more properly, degrees of freedom) only up to 35 or so. Any sample size above that, and one was more or less free to use the Z-distribution's 1.96 as the barrier between "significant" and "insignificant" estimates. Even if a sample of 167 limited our ability to explore the heterogeneity of the effect of bad air on house prices, at least we had a nice estimate of the average effect. Life was simple, and so was the analysis.

We know better now. We knew better then, too, of course, but we lacked the means and opportunities to do better. Four related developments have allowed real estate (and other economic and social) researchers to vastly improve their analytical techniques and allow for deeper exploration, more meaningful insight, and greater confidence in the conclusions than ever before.

The first development was theoretical. The progress of asymptotic theory for data that is heterogeneous and dependent (both spatially and temporally) alerts us to the dangers of oversimplifying the ability of samples of (say) 167 to produce estimators that behave as their asymptotic distributions would indicate when data is not normal. And the problem is worse for data that is generated by power law and similarly ill-behaved distributions. Real estate is a heterogeneous asset that can exhibit these kinds of characteristics, as demonstrated by Roger Brown in his 2004 work.

The second development is the development of computational capability. This is evidenced both in the increase in storage capabilities, and in the ability to do perform massive computational tasks. The terabytes of storage and great increase in computational speed that are now easily available on desktop machines could only be vaguely imagined forty years ago. And for even more complex and challenging problems, many have access to supercomputers, whose capabilities are for all practical purposes, unlimited.

The third development is software. From rudimentary beginnings, econometric and other statistical software has developed to the point even beginning researchers have been provided the opportunity to perform sophisticated and complex calculations, and in so doing, produce sturdy and robust inference.

And finally, the fourth development, and the subject of this book, is the creation of databases that are "big". How big is big? Pretty big, it turns out. It is not uncommon for academic studies to use over a million data points, and private data providers have access to much more. The current record for published real estate research is probably held by Liang Peng and Thomas Thibodeau, who use 26 million housing transactions in their 2017 study of unit-specific risk. Undoubtedly bigger ones are coming.

And when they do, Kimberly Winson-Geideman, Andy Krause, Clifford A. Lipscomb, and Nick Evangelopoulos will be there to guide their users. With great computational power comes great responsibility, and the chapters that follow provide a comprehensive field guide to the responsible acquisition, care, cleaning, description, and analysis of big real estate databases. There are a lot of issues: legal issues, storage issues, statistical issues, and presentation issues, a number of which make plain that big data is not just bigger versions of "ordinary" databases, but can be different in kind as well. The authors' advice will be helpful for any researcher with any sized database, but their emphasis and concentration on big data in real estate will be particularly helpful for cutting edge research in our field. Real estate research is more important than ever: academic, policy and corporate scientists are more attuned to the special nature of real estate and the important roles it plays in the economy, in people's lives and in financial portfolios. The authors are doing all of us a tremendous service.

Edward Coulson
Professor of Economics and Public Policy
Merage School of Business
University of California, Irvine
Professor Emeritus of Economics
Penn State University
Interim Director, Lied Institute for Real Estate Studies
University of Nevada, Las Vegas

Acknowledgements

We would like to thank Holly Wharton for helping organize this disparate group of authors, and Joe Schulman for editing the final version of this book.

Abbreviations

AIC	Akaike Information Criterion
API	Application Programming Interface
AVM	Automated Valuation Model
BMA	Bayesian Model Averaging
CAMA	Computer Assisted Mass Appraisal
CART	Classification and Regression Tree
DFS	Distributed File Systems
EDA	Exploratory Data Analysis
FTP	File Transfer Protocol
GIS	Geographic Information System
GPS	Global Positioning System
GUI	Graphical User Interface
GWR	Geographically Weighted Regression
HPI	House Price Index
LSA	Latent Semantic Analysis
MLS	Multiple Listing Service
OLS	Ordinary Least Squares (Regression)
PII	Personally Identifiable Information
PIN	Parcel Identification Number
RDBMS	Relational Database Management System
REIT	Real Estate Investment Trust
RMBS	Residential Mortgage Backed Security
SFR	Single-Family Residential
SQL	Structured Query Language
TAV	Tax Assessed Value
TOL	Tolerance Metric
SWM	Spatial Weights Matrix
VIF	Variance Inflation Factor

Contributors

Kimberly Winson-Geideman is a member of the faculty at the University of Melbourne, Australia, where she teaches subjects in real estate finance and investments. Her research history includes a broad range of real estate topics including office property valuation, REITs, brownfields, and historic property valuation. She has been writing about the intersection of real estate and technology for several years. Dr. Winson-Geideman is a fellow and board member of the American Real Estate Society, a former board member of the International Real Estate Society, and a member of the Pacific Rim Real Estate Society. She spends her free time exploring Australia in the company of her husband and three children.

Andy Krause is Principal Data Scientist at Greenfield Advisors in Seattle, Washington, USA. His work focuses on developing new products and metrics aimed at improving the accuracy and transparency of real estate market analytics. Dr. Krause was previously a lecturer at the University of Melbourne, where his research examined Airbnb markets, house price indexes, price-to-rent ratios, and furthering the cause of reproducible research in real estate. When not working on R code, Andy can be found exploring the mountains and waterways of the Pacific Northwest via foot, bicycle, or kayak. His professional website is www.andykrause.com.

Clifford A. Lipscomb is the Vice Chairman and Co-Managing Director at Greenfield Advisors, Managing Director of Bartow Street Capital (Greenfield's investment banking entity), and Managing Director of AVM Analytics (Greenfield's software development entity). His consulting and academic research activities have included economic impact analyses for various non-profit organizations and local homebuilders, fiscal impact analyses for local government entities, a revenue equalization study for local legislators, and case study analyses ranging from construction-related impact fees to the economic impact of Wal-Mart. He is a Chartered Valuation Surveyor and holds a BA in Economics and Sociology from Berry College and a PhD in Public Policy from the Georgia Institute of Technology. He and his wife Amelia have five children.

Nicholas Evangelopoulos is a Professor of Business Analytics at the University of North Texas, USA, where he teaches business statistics, data mining, and big data analytics. His current research interests include text analytics, decision analysis, and applied statistics. His articles appear in *MIS Quarterly, Decision Sciences, Decision Support Systems, Communications of the ACM, Communications in Statistics*, and many others. In 2013 he received the UNT College of Business Senior Faculty Research Award, and in 2016 he was named the 2016–2017 UNT College of Business Professional Development Institute Fellow. Each summer he enjoys spending time on the Greek islands.

Lucy Cradduck is a Senior Lecturer in the Faculty of Law at the Queensland University of Technology. Her research interests include the development of appropriate high-speed broadband/internet access policies and the impacts to property use and users from lack of access. Lucy's doctoral (SJD) thesis – *The Future of the Internet Economy: Addressing challenges facing the implementation of the Australian National Broadband Network* – examined issues from an Australian perspective. Her book – *Individuals, Innovation, and the Internet: Why Access Is Essential* – examined issues from an international perspective focusing on solutions to support individuals. In her spare time she enjoys gardening and supporting Manchester United. She can be reached at l.cradduck@qut.edu.au.

Introduction

The real estate industry has a reputation as a late adopter of new technology. This conservatism is driven in no small part by the fact that real estate is conducted one deal at a time with a strong personal element motivating the process. Networking, personal relationships, and sales prowess have been the proverbial keys to success with some of today's largest real estate firms built on this foundation. Market players with sensitive, private, or proprietary information had a distinct advantage over their competitors and were able to convert that knowledge into value.

In many respects little has changed in the modern era; real estate continues to be an industry based on personal relationships, and there is significant reward for those who exploit information asymmetries in the market. But what has changed is the data that feeds the information component; the volume, accessibility, and transferability of data has opened new markets in the industry, primarily focused on information sharing and advanced market analytics. The propagation of the Internet and demand by consumers of all sorts – homebuyers, prospective tenants, and financiers – has helped whittle away at the once indefensible walls of proprietary real estate data, and new firms have disrupted the market by exploiting nuances that were unimaginable even five years ago. The co-modification of real estate, historically marketed as stand alone assets, through REITs, securitizations, and increased interest from public and private funds has raised the bar in terms of expectations, transparency, and technological literacy. This has forced technological progress on an industry that can sometimes be characterized as resistant, or at least ambivalent, to change. Those with the foresight to adapt early to the changes and recognize that data is money now have the competitive edge.

The valuation arm of the profession was at the forefront of this progress, with tax assessors and real estate appraisers some of the first to use spreadsheets and statistical methods to improve property valuations, paving the way for today's Automated Valuation Models (AVMs). Not long thereafter lending agencies began to realize the value inherent in the decades worth of property, risk, and loan performance data they possessed. As the supply of data grew and became more sophisticated, markets developed to feed the never-ending demand for information. Real estate firms came under increasing pressure adapt or face probable failure. The reverberations of this transformation are still being felt as the search for information and competitive advantage in "big data" and the associated "big data analytics" are now the latest trends in the industry.

Big data, machine learning, and artificial intelligence (AI) have made it an exciting time to be in the field of real estate. New data sources and innovative methods are rapidly entering the field making data analytics one of the fastest growing areas within the sector, but because the transformation has occurred so quickly, firms that do not necessarily have a history in real estate or a specialty in the field are making important decisions about the future of real estate data. The impetus for this book is partly driven on that disconnect and partly by our experiences as educators, collaborators, researchers, and consultants in the discipline. What we intend to provide is a book that comprehensively looks at the rapidly changing field of real estate analytics, from an overview of data sources and data types, through the murky waters of gathering, managing, preparing, exploring, and cleaning large, heterogeneous, and, often, error prone data, eventually through to building models using all manner of new data products. We have taken a practical approach to these issues, providing a case study example that can be replicated using data found in the online host repository for the book available at www.github.com/REAIABook/REAIABook/.

Overview

This book is divided into four parts. We begin in Part I with two chapters looking at real estate data, both traditional and "big". In these chapters we cover the current data landscape – what data is available, who provides it and how it can be accessed. We also look toward the emerging trends in big data analytics to see how this rapidly evolving domain can provide input to real estate analyses and benefit the discipline.

Chapters 3 through 7 make up Part II and cover the process of acquiring, managing, and preparing real estate data. An overview of software tools and best practices in documenting data provenance and building reproducible workflows are also included. Our focus here is the practical details of getting from raw data to data that can be analyzed, including a collection of processes and techniques that are often ignored in other discussions of analytical methods.

Part III, chapters 8 through 12, covers modeling and the analysis of prepared real estate data. In addition to chapters on pricing models and the analysis of unstructured text, we also delve into exploratory data analysis and techniques for identifying discordant values – outliers and/or data errors. This section also provides a reproducible quantitative case study using open source or readily accessible data.

We conclude with two chapters (Part IV) addressing some of the legalities associated with big data and concludes with a few thoughts on the future of big data and complex analytical methods in the real estate discipline.

Case study

Chapters 3 through 11 describe the process of completing a complex real estate analysis, from software selection and data management through to statistical modeling. We operationalize these methods with a practical example in Chapter 12. This case study examines the relationships between City of Seattle housing prices and local crime and citizen sentiment. The case study components are made up of traditional data and big data merged together to create the type of analysis that we believe offers big potential to the real estate discipline.

The case study uses data that is freely accessible to any reader who wishes to duplicate our analysis. We have made these data publicly available (to the extent possible) along with all of the code needed for reproduction purposes at www.github.com/REAIABook/REAIABook. A manual is included on the website that describes how to use the code and data. The vast majority of the analytical workflow (minus some of the topic modeling) is conducted in the R programming language, which is also free to download and use. We hope that by following our own advice (and meticulously documenting every step of a complex analysis) we have been able to provide the reader with a clear path for precisely reproducing our results.

Part I

Data

1 Traditional real estate data

Overview of real estate data

Real estate data are multi-dimensional and take many forms including single-family residential (SFR) property characteristics, house price indices (HPIs), different spatial aggregations of office leasing transactional data, real estate investment trust (REIT) returns, and mortgage rate trends. With recent improvements in computing power, larger, multi-faceted sets of property and other data can now be stored, managed, and analyzed. These large, complex datasets are often referred to by the generic term "big data", and although the phrase is expressed quite frequently, what actually constitutes "big data" is something of an enigma. A recent book by Foster, Ghani, Jarmin, Kreuter, and Lane (2017) suggests "there are almost as many definitions of big data as there are new types of data" (p. 3). The author's choose to narrow in on the definition provided by Japec et al. (2015) – "an imprecise description of a rich and complicated set of characteristics, practices, techniques, ethical issues, and outcomes all associated with data." Others have a more specific view of big data – the White House (2014) defines it in terms of the volume of data collected and processed, the variety of data that is or can be digitized, and the velocity of data that can be obtained in real- or nearly real-time.

To provide some background to this book, we initially cover what most researchers refer to as "traditional" real estate data. These data are most often related to sales transactions (or volume) and can include micro-property data (property listings, property characteristics [e.g. tax assessed value (TAV), structural characteristics, geographic identifiers like latitude and longitude], and building permits) as well as macro-housing data (e.g., HPIs, other longitudinal data, measures of supply and demand for leased space, and property data at a more macro geographic scale). Generally, we discuss property data in two broad categories – commercial real estate and residential real estate. This chapter focuses on the different government and industry data sources within these categories that are most commonly used in market analyses. To conclude the chapter, we discuss some issues related to data licensing, which is how the rights to license big data sources are commonly obtained by firms for their own internal research purposes or to create derivative products.

The four *S*'s of property data

There is no official or agreed upon typology or system for classifying the ever-increasing stock of property data that is available. In this book we propose that property data can be organized by determining how it falls across four different dimensions: – Scope, Substance, Source, and Sphere – the 4 *S*'s of property data classification. Briefly, the 4 *S*'s encompass:

1 Scope: The relationship between the intended use of the data and the property industry
2 Substance: The phenomena represented by the observations in the data
3 Source: The origin of or the authority in charge of data collection and dissemination
4 Sphere: The public or private-ness of the data content and data ownership

Scope

The scope of property data refers to the nexus between the original purpose for which the data was collected and its use in the property industry. To be more specific, there are three varieties of scope: Specific, Intentional, and Collateral. "Specific" data are data primarily collected for (and usually by) the property industry itself. "Intentional" data are data collected intentionally for the purpose of third party analysis, but not collected specifically for the property industry. "Intentional" data has a much wider usage. Finally, "Collateral" data are data that results as a by-product of some other action, such as a web search, social media post, or action not directly intended as a data collection exercise.

These three 'scopes' of property data map well onto the Winson-Geideman and Krause (2016) typology of Core, Static Spatial, and Peripheral real estate data. Data with a "specific" scope relates to the "core" data category. "Core" real estate data includes financial data, physical data, and transactional data. According to Winson-Geideman and Krause (2016), financial data includes REITs (e.g. data center REITs, commercial property REITs) and real estate-related stocks. Second, physical data include information about the hard real estate asset, whether that is an unimproved land parcel or an improved property with several structures on it. Third, transactional data include real estate purchases, mortgage origination data, lease data, expense data (included real estate taxes), as well as general economic returns for a single real estate development/investment or a portfolio of investments. With the physical and transactional data types, a common identifier such as property address or other parcel identification numbering system is often used to link these data types together. Examples of combining these two data types include data captured and maintained by property tax assessors and multiple listing services (MLSs).

"Intentional" data relates to Winson-Geideman and Krause's "Static Spatial" category. Within this typology, we consider the most appropriate way to incorporate various types of geographic information systems (GIS) and spatial data. Prior to GIS, these externalities were part of the tacit knowledge of real estate professionals that were analyzed qualitatively, whereas property characteristics, financial returns, and other traditional numeric data were analyzed quantitatively. GIS and spatial data are not financial, nor are they physical or transactional *per se*. Instead, they are "extra-locational." With the innovations made in GIS technologies in the 1990s, real estate professionals and researchers could efficiently use new kinds of data. GIS innovations allowed for extra-locational spatial data to be collected, quantified, and analyzed.

We use the term "extra-locational" to indicate data on spatial phenomena outside the physical boundaries of a property. Examples of extra-locational data include neighborhood information from the Census Bureau, traffic patterns, analyses of proximity to amenities and disamenities (to measure externalities), viewsheds, accessibility metrics, etc. In other words, this data type "quantifies how the property itself relates to existing external physical realities" (Winson-Geideman & Krause, 2016). Extra-locational data, which has

Table 1.1 Data scope typology examples

Property-Based	*Extra-Locational*	
Specific/Core	*Intentional/Static Spatial*	*Collateral/Peripheral*
Sales Transactions	Census Bureau Data	Internet Searches
Lease Transactions	Road Network Data	Transit Ridership Data
Mortgage Data	Geographic Data	Live Traffic Data
Tax Assessment Values	Aggregated Spatial (Core) Data	Point of Sale (POS) Data
Property Level Data (PLD)	Urban Planning Forecasts	Geo-Located Tweets
REIT/Real Estate Stock Data	Spatial Economic Indicators	Pedestrian Traffic Counts

been used for nearly two decades, often remains somewhat limited in terms of temporal resolution (i.e., is not often available in real-time, etc.) and usually possesses a well-defined spatial extent (e.g., Census tract, Census block group, school district).

Finally, data that is "Collateral" in scope maps to Winson-Geideman and Krause's "Peripheral" data. Collateral data refers to data whose collection is incidental to or a by-product of some other process. Collateral data are varied, disparate, and often available in real-time. It may be useful to think of Collateral data as being primarily human-focused whereas Traditional and Intentional data are typically non-human in nature. Collateral data are often remotely-sensed (gathered mechanically) instead of collected directly by real estate professionals or government workers. Finally, Collateral data are increasingly being used in real estate predictions and forecasting.

To summarize our conception of real estate data in the context of big data, Table 1.1 (based on Winson-Geideman and Krause (2016), 1) provides a non-exhaustive list of the three types of real estate data used or practically available to use in real estate research and analysis.

Substance

The substance of a particular set of data refers to the real world phenomenon that the data represent. There is no bounding set of what the substance can be, especially in data with a collateral scope, but the most common substances are: economic, physical, location, and human. Economic data refers to those data that represent information on economic or financial transactions. In short, it includes any situation in which money is exchanged or values are represented in relation to real estate. Physical data directly describes the hard, real asset that underlies real estate. This can be the structure or the land. Location data provides an orientation to the actual position of the asset (direct scope) or an externality or other party (intentional or collateral scope) on the earth or in relation to some fixed or known point. Location data also includes information regarding the boundaries of various administrative, natural, or market areas, or even the right-of-ways or routes of thoroughfares, transit lines, or waterways. Finally, human data digitally represents the activities of one or more people. This can be the act of walking, driving, spending, resting, tweeting, searching, calling – anything people do that leaves a digital footprint.

Some data may have more than one substance. A set of data giving the geo-located points from which people made mobile device searches of real estate listings has both human and location substance. If the same dataset included, for example, the actual booking of an Airbnb lodging, then it could also be considered economic. While most, if not all, combinations of scope and substance are possible, some are certainly more likely than others. Often physical and economic substance data is of a direct scope, while location data is primarily direct or intentional in scope. Human data is most often collateral in scope.

Source

There are three primary sources of property data: private industry, government, and academia/non-profit organizations. We consider government to be separate from academia and the non-profit sources because in most cases, government data is collected and distributed by mandate, whereas data from academia and non-profits are usually, though not always, collected and distributed on an ad hoc, or occasional basis.

The source of the data is important in understanding the frequency with which it is updated, the cost of accessing the data, and likely standardization (or lack thereof) of the data product. Government data and industry data are more likely to be produced on a regular schedule and distributed in a standardized, at least internally, fashion. Industry data, however, most often comes with a cost, whereas much, but not all of government data is free or open for use. Data sourced from non-profits or academic sources is often distinct to a particular project or cause and may not be updated regularly or standardized to integrate well with other datasets. Additionally, academic/non-profit data is usually some combination of government and industry data. As a result, we focus on government and industry data below, reserving a small discussion on academic and non-profit data resources near the end of the chapter.

Sphere

The sphere of data refers to whether the content is public or private. For example, data about the street network has a public sphere in terms of content. Conversely, data about the structural characteristics of a single family home has a private sphere in terms of content because a home is private property. Both public and private content can come from a government (public) or an industry (private) source. The street network data may be owned (and distributed) by a government (public) or by Google (industry). Likewise, the private content of a home's structural characteristics can be owned and distributed by an industry entity, such as the Multiple Listing Service, or by a public entity, like a local government assessor or valuer's office.

Across the four dimensions of the data, any combination of *S*'s is theoretically possible. There can be specific (scope) physical (substance) industry (source) private data, or collateral human government public data, and all manner of permutations in-between. Some combinations are certainly more likely than others, but we do not present an entire list of those possibilities and their propensity[1] here.

Note that we have not explicitly addressed the issue of data access or cost in our four dimensions of data. We have omitted this aspect of property data for two reasons. First, the costs to access data are often different for different users or potential users. What may cost thousands of dollars for a private company to access may be free to government or non-profit researchers, or companies may share data with each other, but charge governments for access. Second, the accessibility of various data sources is always changing and examples of one or the other may be well outdated by the time this book is published. In regards to access, the reader or potential analyst should keep in mind that some data has a cost, and occasionally in property that cost can be very steep, but that the cost may vary given the analyst's own position. More on the legal nature of data access and sharing is presented in Chapter 13.

Data examples

With this background, we set out to describe some of the more frequent sets of traditional data sources that are commonly used in real estate research and analysis. Within this chapter we focus on sources that have either a Specific or an Intentional Scope. A closer look at big data, usually Collateral in Scope, follows in Chapter 2. We split the data by source here, first looking at government sources and then private industry. Within each of these groupings there are a variety of different Substance and Spheres represented. As these two dimensions of property data are highly variable, we do not attempt to classify the data below by them at this time. Finally, as non-profit and academic sources are highly idiosyncratic, we discuss at the end a handful of the most consistent and well-recognized data that come from academic or non-profit sources, regardless of scope.

Specific scope, government source

Key government sources of data are local jurisdictional agencies that collect real estate data to facilitate valuations for tax purposes. Tax assessors or valuers determine the value of all taxable properties within a designated county, municipality, or other specified geographic area. In the United Kingdom, valuations are governed through the Valuation Office Agency and conducted at the council level; in the U.S. they are conducted by county agencies, and in Australia land rates are determined by the local councils. Many places offer a free online database of assessed property values, which sometimes include actual sale prices. Other property and transactional characteristics are also included, but this varies widely by jurisdiction.

The collection of data for tax purposes is inconsistent among states in the U.S. as well as among different countries. This is primarily due to the variety of regulations that govern privacy and disclosure as well as how (and at what level) various real property taxes are levied. For example, in 12 of the U.S. states[2] the sale price of a property transaction is not mandatorily made public. Lease transactions, both residential and commercial, are even less likely to exist as publicly reported data, thus there are few government sources that provide income information for leased properties. Overall, the information governments are likely to possess and potentially distribute is limited to the physical properties within their jurisdictions, and depending on the locale, some agencies may also maintain databases of sales transaction prices, while lease information is much sparser.

Other government agencies such as the Federal Housing Finance Agency in the U.S., the Reserve Bank of Australia (RBA), the Bank of England, and the European Central Bank collect and monitor the state of the broader housing market. These agencies collect and often distribute information on sales volumes, mortgage rates, house price indices, new housing starts, overall borrowing on real estate, and other metrics that indicate the contribution of the real estate industry to the broader economy. The availability of data varies by country, as different federal government are required to undertake different reporting mandates and policy evaluations.

In the U.S., the Department of Housing and Urban Development (HUD) facilitates housing opportunities for low- to moderate-income and first-time homebuyers through government programs. As such, approximately 15% of houses sold in the U.S. are considered "HUD houses," or homes that have a government-insured (FHA) loan. The list of HUD homes is publicly available, as is the list of homes that have been foreclosed on that have a government-insured, FHA loan. Similar programs in other countries include the Australian Department of Social Services and Canadian Housing Mortgage Commission.

Intentional scope, government source

A variety of other government agencies and programs gather data that is used by the property industry but is collected for purposes outside of the confines of the industry. The most common of these are the periodic censuses undertaken at great cost by federal governments around the world. Additional federal agencies, as well as state and local departments, are also active in the collection and distribution of intentional data.

Census

Most countries conduct a periodic census to gather information about the population and plan for future resource allocation, providing an abundance of demographic information that has become essential to real estate research. In the United States, the population census is conducted every 10 years, but the Census Bureau also compiles various housing data in its American Housing Survey (AHS) every other year. AHS data includes home type and size statistics, as well as home values, average rents and mortgages, vacancies, new construction, multi-family housing, and financing characteristics. Most U.S. Census Bureau data are free and publicly available with some exceptions for highly confidential or personalized information.

The Australian Bureau of Statistics conducts the Census of Population and Housing every five years. While participation is compulsory, some questions, such as religious identification, are optional. Information ranging from occupation, household size, income, transportation, and housing is collected from households and released publicly at various levels of aggregation. Strict privacy laws prevent individual or small group information from being released in the fear that it may be subject to de-anonymization.

The census in the U.K. is conducted every 10 years by various jurisdictions, a process that dates back to 1801. While most information is released in aggregated form, Samples of Anonymized Records (SARS) that include individual, anonymized records were made available in 1991, 2001, and 2011 and are often used by social scientists and others in research.

Other government agencies collect data to better understand population characteristics, economics of cities and regions, traffic and commuting patterns, changes to environmental conditions, public safety issues, and other metrics of social function. Because real estate is heavily influenced by the impacts of externalities, data from all manner of government departments is often used to aid in better understanding the operations of the market.

Specific scope, private source

Much of the specific data about real estate is collected and disseminated (sold) by private industries. There are three types of industry players in the data space: (1) agents or marketers; (2) real estate service providers, advisors, and consultants; and (3) data aggregators.

Agent-centric or marketing firms are either engaged in the business of directly selling or leasing of real estate or in offering a platform for others to do so. The MLS in the U.S. is a long-standing example of a data source that acts as a central hub for agents looking to sell their properties as well as a data collector that sells the information from the agents to other parties such as valuers. A new business model that relies on agents bidding for space as well as traditional online advertising has driven the rise of firms like Zillow in the U.S. and Domain in Australia. These sites provide information on the property market (usually just residential) for free to consumers, driving revenue through ads and agent fees. These firms have a great deal of data on specific properties as well as broader market trends, some of which is available for download. CoStar, Real Capital Analytics, and LoopNet are operators in the non-residential space.

Service providers and advisors make up another core component of the private data landscape. These are firms that specialize in advising clients on market actions, and in the course of their duties also collect and distribute data. Many of the larger firms in this space are global, such as CBRE and JLL, and employ research teams that focus on analysis and data generation. Much of the data available from providers and advisors comes in a two-tiered system. There are freely available reports and periodic data releases that offer aggregated and summary data of various market segments. This information acts as a marketing tool to attract subscribers and paying clients who then have access to better and more directed information.

Finally, the third group of players in this realm is data aggregators. There are two types of data aggregators, those whose main business is the collection, standardization, and dissemination of property data, and those who do so as a consequence of other activities. CoreLogic and Black Knight represent two of the largest data aggregators, whereas others such as Attom Data Solutions (the parent company of RealtyTrac) reconcile multiple data sources into a "superset" of real estate data that it provides to its customers (Lipscomb, 2017, p. 32).

Appraisal management companies often fall in the latter group. Appraisal management companies in the U.S. act as an administrative liaison, or middleman, between lending institutions and real estate appraisers. AMCs order a real estate valuation when a mortgage lender is considering extending a loan to a buyer, and once the appraisal is completed, the AMC will deliver the appraisal report to the lending institution.

Valuations in Australia and the U.K. are conducted somewhat differently, without the AMC equivalent. Lenders in Australia engage directly with valuation companies whose employees have completed the necessary qualifications to produce formal valuations (not appraisals). Appraisals in Australia, incidentally, are considered informal estimates, are not legally binding, and are often issued by real estate agents with knowledge of the surrounding area. Valuers gain access to comparable data from CoreLogic RP Data, or property data delivered through Real Estate Institute of Victoria (REIV) and compiled through a network of property valuers around Australia.

Valuations in the U.K. are governed by the Royal Institute of Chartered Surveyors (RICS) and are conducted by property valuers, also known as valuation surveyors. Commercial property valuers source their comparable data from companies like Property Data in partnership with Real Capital Analytics or Landmark Analytics for residential and other types of property valuations.

Intentional Scope, private source

Organizations that represent the intentional scope, private source framework focus on the identification, development, and dissemination of very specialized types of information. Mapping companies such as Pictometry, NearMap, and ESRI collect spatial data through aerial photography, photomapping, and topography analysis and often provide GIS services to their clientele. Demographic firms develop profiles for neighborhoods or other geographic areas for marketing purposes, while planning and resource consultants develop databases for the purpose of organizing future transportation routes, preserving natural resources, or affordable housing provision. Most of these datasets are proprietary, although some companies like Walkscore.com provide user-friendly interfaces for web access without charging a subscription fee.

Non-profit or academic data

Universities, particularly those with large, research-oriented faculties, often house one or more research centers focused on a specific discipline or unifying theme. Some of them collect primary data which is then used in combination with other sources of data to thoroughly evaluate a subject. The University of Michigan Consumer Sentiment Index, for example, is a well-known and oft-cited survey that regularly quizzes a sample of 500 respondents on their attitudes toward business, personal finance, and spending. The inflation expectations and other economic measures that are derived from the survey are regularly used in real estate research. Other centers warehouse data from a variety of sources in an effort to make it easier to access and therefore use. The Australian Urban Research Infrastructure Network housed at the University of Melbourne does just that, bringing in data from government as well as private enterprises through contractual arrangements.

Some research centers lodged in universities are focused less on issues of national importance and more on understanding the factors that influence cities or regions. Since 1971, the Real Estate Research Center at Texas A&M University has been publishing a number of different reports related to the condition of the real estate markets in Texas. Similar centers at the University of Washington, University of Cambridge, and the University of Pennsylvania, among others, support regional as well as national research initiatives.

Non-profits and think tanks also provide a great deal of insightful, data-driven research on real estate and community-related issues. While many operate independent of any government or corporate entity, others represent a particular industry or special interest and are supported through a combination of membership fees, small donations, and/or large-scale philanthropy. Many support outside research through competitive grant processes. The Urban Land Institute (ULI), for example, has represented the land development industry for almost 90 years, providing research and education on broad issues such as the environment and the economy to more local issues such as infrastructure provisions and public policy. The Lincoln Institute of Land Policy in the U.S. focuses on evidence-based research into land policy, offering free property value, tax, and other databases to researchers, and the Grattan Institute focuses on policy, transportation, and education issues affecting Australians. Others, such as the Overseas Development Institute in the U.K., promote research into policy issues with a global perspective.

Data licensing

Data licenses are the most common way to convey certain rights to end users of data. There are three areas common to most data license agreements: Delivery, Term, and Fees; Permitted Uses/Applications; and Permitted Users. While other items are included in data licenses because of their contractual nature (e.g., signature pages, remedies in the event of breach of contract, definitions of key terms), here we discuss only those items we believe directly relate to the conveyance of rights to use real estate data. Below we discuss each of these items.

Delivery, term, and fees

"Delivery" refers to when data will be available to the customer, whereas the "term" of the license typically includes the dates the license is active as well as instructions on what to do at the end of the license period if the customer wishes to renew. A set period is often covered by the initial data license (or agreement) and optional extension periods incur an additional fee. The Delivery, Term, and Fees sections of a data license follow a format similar to the examples provided below.

Delivery

> Example 1: *Company* shall deliver a one-time load of Licensed Information to Licensee no later than FIFTEEN (15) business days after the Effective Date. Thereafter, no updates shall occur. All Licensed Information shall be in the form of an ASCII delimited text file, and will be delivered via a secure, customer-dedicated folder on *Company's* file transfer protocol server.
>
> Example 2: *Company* shall provide Customer with the Services listed via the specified delivery method.

Term

> Example 1: This Schedule shall be effective for an initial Term beginning on the Effective Date and ending once the Action has settled, at which time Licensee must (1) notify *Company* of the fact that the Action has been settled; and (2) destroy the licensed Information in accordance with the Agreement.
>
> Example 2: The term of this Scope of Work is for 2 years, commencing on the Scope of Work Effective Date. Thereafter, the term will automatically renew for additional successive 12-month terms. Either Party may forego automatic renewal of this Scope of Work by giving the other Party at least 60 days' written notice of termination prior to the expiration of the then-current term.

Fees

> Example 1: Customer shall pay to Company the Fees set forth. Customer shall be solely responsible for the payment of all fees and expenses incurred under this Scope of Work, and *Company* acknowledges that its rights and remedies with respect to the Services provided to Customer shall be solely against Customers.
>
> Example 2: In consideration of the license granted herein, Licensee agrees to and shall pay *Company* a flat licensing fee of *XXX ($)*.

Permitted uses (or applications)

Another common section in data licenses is Permitted Uses, sometimes referred to as Permitted Applications. This section often outlines what is and what is not permissible to do with the licensed data. For example, this section may indicate that a customer can use the data for its internal research purposes, but cannot create a derivative product from the data. A derivative product is a term used to describe a product produced and sold commercially by the customer using the licensed data. While some data licenses permit the creation of a derivative product, often those data licenses are more expensive or a certain percentage of the gross revenues of the derivative product must be paid to the company providing the licensed data. Below are a few examples of Permitted Uses.

Permitted uses

1 Licensee may use the Licensed Information internally for research, reference, valuation modeling, and analysis.
2 Licensee may sublicense, disclose, and distribute its analysis to its single Customer, along with underlying data in support of such analysis.
3 Licensee may incorporate and store the Licensed Information within Licensee's internal systems necessary to perform the Permitted Uses.

Permitted users

The last section commonly found in data licenses that we will address here is Permitted Users, a section of the contract that identifies who can use the licensed data. Many distributors are quite explicit regarding who can use the product, so they almost always obtain a list of permitted users from the customer. The fear is that the larger the number of permitted users, the greater the risk that data may be used for a purpose not allowed in the data license. The company expects to be compensated for the increased risk that the data could be inadvertently used to create a derivative product or end up in the possession of a company that did not originally license the data. Therefore, the more permitted users there are, the higher the licensing fees. An example of a Permitted Users section of a data license is provided below.

Permitted users

Customer and approved client shall use the Services for their own internal business purposes of due diligence and for both internal and external use solely on behalf of Approved Clients in connection with litigation or other disputes relating to alleged untrue or misleading statements made by issuers, brokers-dealers, originators and other parties involved in issuing securities purchased by Approved Clients. Neither Customer nor Approved Clients shall resell, relicense or redistribute the Services in whole or in part. Approved Clients, as well as their experts, vendors, and agents, shall be considered Permitted Users of the Services.

Notes

1 With the rate at which data is created, the propensity of each combination is likely not known and possibly practically unknowable at any given point in time.
2 Alaska, Idaho, Kansas, Louisiana, Mississippi, Missouri (partial), Montana, New Mexico, North Dakota, Texas, Utah, and Wyoming. Source: www.zillowgroup.com/news

References

Foster, I., Ghani, R., Jarmin, R., Kreuter, F., & Lane, J. (2017). *Big Data and Social Science: A Practical Guide to Methods and Tools*. Boca Raton, FL: CRC Press.

Japec, L., Kreuter, F., Berg, M., Biemer, P., Decker, P., Lampe, C., Lane, J., O'Neil, C., & Usher, A. (2015). Big data in survey research: AAPOR task force report. *Public Opinion Quarterly*, *79*(4), 839–880.

Lipscomb, C. A. (2017). The next generation of AVMs. *Fair & Equitable*, March, 29–33.

White House. (2014). *Big Data: Seizing Opportunities, Preserving Values*. Washington, DC: Executive Office of the President.

Winson-Geideman, K. & Krause, A. (2016). Transformations in Real Estate Research: The Big Data Revolution. *Presentation at the 22nd Annual Pacific-Rim Real Estate Society Conference*, Sunshine Coast, Queensland, Australia, January 17–20.

2 Big data

What is big data?

Big data is a trend that has captured the attention of businesses, governments, and academics alike in disciplines as diverse as finance, healthcare, computing, and transportation. Businesses use it to define market strategies, improve point-of-purchase sales, and precisely target advertising. Governments use it in urban planning and law enforcement, while academics are taking advantage of it to promote, define, and expand research agendas. With corporations now collecting petabytes (one million gigabytes) of data, users expecting instantaneous results, and new technologies required for analysis, big data has created many opportunities for those willing to harness its potential. Big data crosses borders and has the capacity to disrupt the way we think about research in economics, the social sciences, medicine, and business for years to come. Some have likened it to Henry Ford's system of mass production that set the standard for technological progress over a century ago (Boyd & Crawford, 2012). It is arguably the standard for technological progress of our time, with the potential to provide greater degrees of insight into human motivations, actions, and behaviors than anything has before.

The seemingly rapid accumulation of data (not to mention the endless supply) has overwhelmed some researchers as they seek to recalibrate old methods and apply them to new data. Even the task of identifying big data is fraught with problems. Many users have adopted the "you know it when you see it" approach, a not unreasonable method when one considers the changes taking place in the big data realm. Others have doubled down on one of the numerous definitions that exist, be it general or very specific in nature. Even though some may claim to be working with big data, it is not necessarily clear that they are when the different definitions are considered. That said, a number of useful and descriptive definitions have arisen since that serve as a framework for most of the big data research that is being conducted around the world. Corporations that collect the data, researchers that analyze it, and governments that use big data have generated these definitions.

So, what exactly is big data? Madden (2012) provides a simple definition, describing it as data that are "too big, too fast, or too hard for existing tools to process." Others refer to it as sets of data so large and complex that they require new or improved computational tools for proper analysis (Burkholder, 1992). The *Financial Times* Lexicon refers to big data as "the term used to describe the huge volumes of data generated by traditional business activities and from new sources such as social media."[1] Oracle takes a results-driven approach, classifying big data as "the derivation of value from traditional relational database driven business decision making, augmented with new sources of unstructured data," while Microsoft refers to it as a term that is "increasingly used to describe the process of applying serious computing power – the latest in machine learning and artificial intelligence – to seriously massive and often highly complex sets of information" (Ward & Barker, 2013).

Gil Press, writing for *Forbes* in 2014, identifies 12 definitions, some serious such as the widely cited McKinsey & Co definition, "datasets whose size is beyond the ability of typical database software tools to capture, store, manage, and analyze," and others not so serious, such as the merger of Madame Olympe Maxime and Lieutenant Commander Data (characters in Harry Potter and Star Trek, respectively). Press ends his piece by coyly asking "what's yours?", suggesting the need for a cohesive, straightforward definition of big data on which reasoned debate can be based.[2]

Technical characteristics of big data

While each of the aforementioned definitions differ to varying degrees, there are several features that tend to arise repeatedly and thus are recognized as the basic characteristics of big data. It is from this framework we form the definition used throughout this book. Commonly known as the 3-*V*s – volume, velocity, and variety – the identification of these characteristics are typically accredited to a white paper written by Doug Laney, an analyst at META Group (now Gartner) in 2001. In that paper, Laney noted that the new forms of data are characterized by immense size or volume, constant streaming or velocity, and the variety of the forms they take, both structured and unstructured. Laney specifically defined big data as "high-volume, -velocity and -variety information assets that demand cost-effective, innovative forms of processing for enhanced insight and decision making" (Laney, 2001).

Discussions of big data typically begin with volume, or the scale of the data. As is implied in its name, big data is not only big, it is *really, really big*. With IBM estimating that 2.5 quintillion bytes of data are created on a daily basis and that 90% of what exists today was collected in just the past two years, it is clear that the sheer amount of data that has and is being collected presents a wide range of challenges to stakeholders.[3] One of these challenges is determining how much data qualifies as "big." Intel categorizes the data collected by individual companies as big when they are "generating a median of 300 terabytes (TB) of data weekly" (Ward & Barker, 2013). But is assigning a quantifiable parameter necessary or desirable when categorizing data, especially when that target is likely to vary widely due to a number of independent factors? Is a lot of data "big" data? The generally accepted answer to both questions would probably be "no," specifically when one considers the other characteristics or "V's" that constitute the big data definition we use here.

The velocity of data refers to the speed and frequency that the data are generated. Data that is "big" is constantly streaming, updating, and being created in such a manner that it is nearly impossible to contain. Consider Twitter, for instance. With a worldwide user base (79% outside of the U.S.) and 310 million monthly users,[4] approximately 6,000 tweets are being created each second, or roughly 350,000 per minute.[5] The website Internet Live Stats shows a running count of the approximate number of tweets produced throughout the day, an excellent visual representation of the velocity at which new data are being created. When multiplied by the number of private enterprises, governments, and educational institutions that are also producing constantly streaming data, it is clear that big data is a frontier with many avenues to explore.

Data that are "big" are also characterized by the variety or various configurations of information that are created. The variety of data falls into two general categories: structured and unstructured. Structured data are data that follow a predictable pattern or set of rules. Examples include dates, amounts expressed in currency, and addresses. Structured data are generally easy to collect, store, and understand, especially when compared to unstructured data. Unstructured data are data that lacks conformity and includes items such as photos, videos, web links, and various combinations of characters. Each of these types of unstructured data has different storage, scaling, and analytical requirements, which can make them difficult to manage.

In the years since Laney's three *V*'s definition was developed, others have contemplated the addition of other characteristics such as veracity (accuracy or dependability),[6] validity (logical inferences), value (is it meaningful, useful or valuable),[7] and volatility. While each of these additional *V*'s are worthy of consideration in any serious study involving big data, they should be considered in studies using *any* type of data be it big, small, conventional, or otherwise. Laney's three *V*'s are the characteristics that distinguish big data from other types of data and give it its unique and cutting edge appeal.

Big data sources

Big data come from a variety of sources. They are literally everywhere and you are probably helping create big data, even though you may not know it.

Most discussions of the sources of big data begin with social media. Many are familiar with Facebook, Twitter, LinkedIn, and Instagram, and to a lesser extent Snapchat. All perfectly embody the three *V*'s by producing incredibly large amounts of data around the clock in numerous formats, and all follow the same basic structure – free access to individuals and institutions in an effort to build their user base, with the option of paid, targeted advertising receiving priority placement in news feeds. Frequent users of the sites will likely encounter advertisements from local real estate agents, construction and renovation companies, community groups, and others looking to promote local business.

Not only are these sources used as a marketing medium for home, land, and commercial ventures, they also serve as a portal for opinions, prejudices, and biases. Information from public profiles can be searched, downloaded, and used to evaluate promotional materials. Facebook and Instagram users have the option of commenting on posts or clicking on emoticons representing predetermined content such as "like," "love," or "wow." Facebook Insights permits users to track different metrics about page performance, identify the most engaging posts, and determine audience availability. Twitter users use hashtags prior to words or phrases to identify reoccurring themes or trends such as #academyawards, #interestrate, #houseprice, and #foreclosure. Some of these topics trend worldwide, while others capture a more localized audience. Tweets that are geocoded can be used to analyze community response and expectations to a specific event such as a neighborhood festival or a rash of crimes. Twitter Analytics allows users to track the number of profile visits, followers, and the impressions other users have about their tweets.

Search engines such as Google, Bing, and Yahoo! also provide a wealth of information in the big data context, predicated on the assumption that internet searches represent the interests, events, and daily experiences of users. Companies recognize the benefits of ensuring that their website is listed first in a generic search, but beyond that, they also recognize the value inherent in the statistics that are generated. Google creates a record each time a search term is used, then feeds it through an algorithm that logs information about that specific search term including time, place, and frequency. The data are aggregated and released publicly through Google Trends.

While interesting and cutting edge, studies incorporating data acquired through Internet searches are not always accurate. Probably the most famous (and notably debunked) study using Internet searches is the case of "Google Flu." In a paper published in the journal *Science*, researchers noted that Google searches related to the flu outperformed the Center for Disease Control's predictive models by two weeks. The results, they suggested, could be replicated in subsequent years to better anticipate and locate flu occurrences with potentially life-saving effects. Unfortunately that potential failed to materialize, and in 2013 the model missed the peak of the season by 140% (Lazer & Kennedy, 2015).

More traditional sources of data are now sources for big data as well. Property prices, descriptions, and characteristics are constantly being changed, added, and removed from the MLS in the United States and similar sites in other countries. Websites such as Realtor.com, Trulia, Zillow, and Redfin (U.S.), Domain and RealEstate.com.au (Australia), as well as Rightmove, Zoopla, and some alternatives in the U.K.[8] serve as search engines for their respective national housing markets. Even eBay permits property to be sold on its website. Data aggregators, including CoreLogic and Black Knight, compile, standardize, and sell the raw big data necessary for these firms to improve their day-to-day products.

Government sources range from information collected about mass transit usage, healthcare, public services, and education. Big data is used to improve service delivery by better anticipating where, when, and how much service will be needed in a given location. Private companies are using this information to provide additional value to clients. Windemere Real Estate, for example, provides potential buyers with detailed commuting information using road network and traffic data (Sonka, 2014) as well as a comparative market analysis.

Financial data related to publicly traded corporations such as banks, real estate companies, and REITs are generated globally and around-the-clock. The terabyte of information that is collected from the New York Stock Exchange (NYSE) each day, for example, is fed into algorithms that support high frequency trading (HFT), with investors relying on its nearly instantaneous results to gain a competitive edge. While margins are generally small, trades are quick and profitable due to their frequent nature. The practice is not without risk, however. HFT is often cited as the catalyst for the Flash Crash of May 6, 2010, captured in Michael Lewis (2014) book *Flash Boys: A Wall Street Revolt*, when large amounts of equity were withdrawn from the market, exposing investors to significant losses in a matter of seconds.

These examples are just a few of the many ways in which big data can transform the way we think about research. What makes it interesting and valuable is that much of it is generated from what are often considered normal, everyday tasks – logging on to Facebook, searching online for a particular shop, and using swipe cards to access public transportation, a building, or a bank account. Each of these actions creates a record, a record that then becomes part of a much bigger database of activities and behaviors that are constantly occurring in many different places. And with the advent of blockchain technology (i.e., distributed consensus model described in Crosby, Nachiappan,

Verma, and Kalyanaraman [2015]), the confirmation of the accuracy of these records can be enhanced.

While appealing at some level, the rapid inundation of data means that researchers need to exercise serious caution when selecting what will be used in their studies. Storage and computing issues plague big data, making the rigorous assessment of it sometimes more difficult than conventional data assessment. It is important to keep in mind that data are not valuable because they are big. Data are valuable because they give insight into a specific question, a question that has relevance in a particular context. If that simple standard is not met, then the data are irrelevant.

IT architectures and big data storage

The *V*'s of big data have a significant impact on how users – governments, businesses, and researchers – collect, manage, analyze, and visualize data. Throughout this book we use the term "manage" to describe the processes of storing and accessing data. Changes made to data to ready it for analysis (i.e., data preparation, exploration, and cleaning) are referred to as "munging." Each of these procedures is important to the research process, and the introduction of big data has produced a number of challenges, especially when it comes to managing the volume and flow of the data that are created.

The proper management of big data requires tradeoffs between speed, efficiency, and access. Because the needs of the particular enterprise and its users are paramount to extracting the value inherent in the data, big data management systems typically focus on one of the factors, sometimes to the detriment of the others. An enterprise that requires immediate and constant access to real-time data, quick analysis, predictive modeling tools, and data queries will choose speed over efficiency and access. Conversely, a company that only utilizes historical datasets occasionally and is concerned with maximizing storage space (memory) and minimizing energy consumption, may prioritize efficiency. Finally, those with highly sensitive data may need to sacrifice some measure of speed and efficiency to balance information retrieval with security concerns, thus they prefer systems with access limitations.

Out of these tradeoffs have arisen two primary types of big data storage systems: data warehouses and data marts. Data warehouses predominate in situations where data is not needed immediately or at a rapid speed. Often, the data stored in a data warehouse does not yet have an identified purpose; the warehouse manager is simply filing away current data for some yet-to-be-identified future use. Data for which current demand is low and/or slow is often referred to as "cold." Cold data are usually stored in data warehouses that focus on efficiency as opposed to speed. Much of this efficiency is gained by storing data on large and slow (but cheap) optical disk technology. The level of security in place in data warehouses will depend upon the nature of the data.

Data for which the current demand is very high is considered "hot" with data marts representing the best storage option. Hot data are usually kept on fast but expensive flash memory devices. Data mart storage options allow for low latency and high performance access. Real-time predictive analytics and/or data with thousands of simultaneous users will nearly always be stored in some form of data mart. Hot data storage is typically expensive to maintain and access, and it often is much more difficult to control, thus making it less secure.

A related issue here is the physical location of the data storage. Three solutions have arisen: on-site, virtual/cloud, and hybrid. Both data warehouses and data marts can use any of these three approaches, depending on the user's needs and financial resources. On-site data storage solutions involve an investment by the enterprise in physical hardware that needs to be installed and managed at one of its locations. These solutions allow for a greater potential to control access and may also help limit external network traffic if the data are hot. Virtual or cloud storage options remove the burden of physical storage apparatus from the user. However, some level of security and general access in the case of external network failure may be forfeited. A hybrid system may be utilized to best optimize a company's data storage, with hot and low-security data hosted virtually and colder, more sensitive data held in-house. In hybrid systems, data may be held in both locations as well, with redundancy providing contingency in the case of failure at either site.

A number of large tech firms offer virtual storage products, often with analytical options (covered later) built in to their packages. These include, but are not limited to: Amazon (S3), Microsoft (Azure), Google (BigTable), HP (Big Data Server), Cloudera (Cloudera Hadoop), and Oracle (Appliance).

Computing architecture

A key characteristic of big data is that it requires specialized analysis as well as storage. While we will discuss big data analytical techniques in later chapters, a more general discussion of the computing architecture is relevant here when covering storage.

Two key advancements have been made in computing architecture with regards to improving big data analytics. The first is parallel computing. Parallel computing means that a given computational task is distributed over multiple processors that run simultaneous computations. Originally reserved for large mainframe servers, even today's laptop computers have multiple processors, and with the proper software they are able to take advantage of parallel computing. Parallel computing can be used to increase the speed at which computations are done; however, memory is shared by the individual CPUs, thus limiting the amount of data being analyzed.

The second advancement is clustered or distributed computing. In this case, a computational problem and/or large dataset is broken into a number of small blocks that are spread across different machines or, in the case of virtualized distributed computing, across different virtual machines that could be all hosted on the same machine/server. The rise of distributed computing has created distributed file systems (DFS). The Hadoop[9] Distributed File System (HDFS) is the leading system in this regard and offers a scalable and fault-tolerant system whereby big data sets are stored and computed in a distributed fashion. Distributed computing can approach problems that are too large (memory- or data-wise) for a single machine to handle. In HDFS, large data files are split into blocks of approximately 128 MB in size, and each block is typically stored in three copies across different nodes of a server cluster. Computational tasks in HDFS are performed following a computing model known as *MapReduce*. In MapReduce, a number of local processes called *mappers* operate on individual data blocks, running initial computations. A *shuffle and sort* process consolidates intermediate data from the mappers and forwards them to a number of *reducers*, which condense the data into a single output. The MapReduce model speeds up processing and minimizes network traffic. It also offers programmers a clean abstraction that hides all parallel processing housekeeping tasks and lets developers concentrate on writing appropriate map and reduce functions that serve a certain computational purpose.

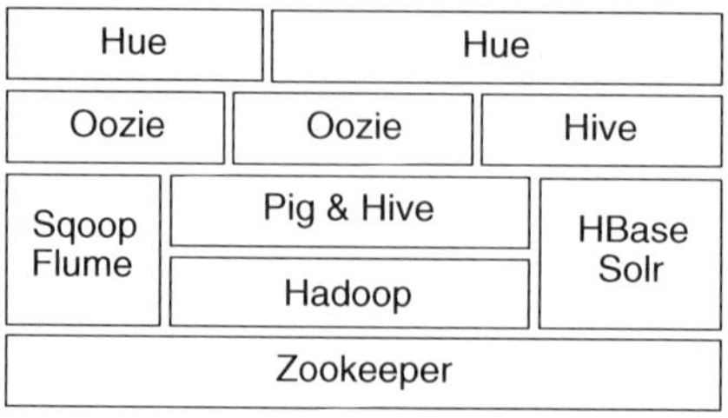

Figure 2.1 The Hadoop Ecosystem as implemented by Cloudera

The Hadoop Ecosystem

In order to make Hadoop more accessible to non-programmers, a suite of user-friendly applications has emerged in recent years. This is known as the Hadoop Ecosystem. A specific version implemented by Cloudera is shown in Figure 2.1. The ecosystem includes the main components listed below.

- *Apache Sqoop*[10] is an application that transfers data between HDFS and an existing relational database (RDBMS). Sqoop automates the execution of all the map-reduce operations needed in the ingestion of the RDBM into HDFS, or the exporting from HDFS to a RDBMS. Sqoop is also used by LinkedIn. Facebook uses Scribe for similar purposes while Google uses Gateway, and Yahoo! uses Data Highway.
- *Apache Flume*[11] is an application that collects data from network sources. Its architecture is based on streaming data flows, with a source, a channel, and a sink as its main components; sources listen for and consume events, channels transfer events from sources to sinks, and sinks provide Flume agents pluggable output capability. Flume is used to efficiently collect, aggregate, and move large amounts of streaming event data such as web logs. Robust and fault-tolerant, it offers a recovery mechanism where standby cluster nodes are ready to automatically take over when the main machine fails. Kafka is a similar product used by LinkedIn and Scribe is used by Facebook.
- *Apache Pig*[12] and *Hive*[13] are data manipulation languages. Pig is a high-level language for expressing data analysis programs, together with the necessary infrastructure for evaluating these programs. The main strength of Pig programs is that their structure supports parallel processing, making them suitable for handling very large data sets. Hive provides an SQL-like language called HiveQL with schema on read, meaning that the data structures (schema) are defined after reading the data, as opposed to the typical RDBMS approach that requires full schema definition before reading the data. Pig and Hive are also used by Facebook, LinkedIn, and Yahoo! and Google uses Sawzall.
- *Apache HBase*[14] and *Solr*[15] are non-relational, not-only-SQL (NoSQL), distributed databases. In Solr each record is called a document. Rather than tables, Solr uses collections of documents that are indexed and searchable, supporting JSON, XML, CSV, MS Word, and PDF formats. HBase is the equivalent product used by Facebook and Yahoo!, Google uses BigTable (HBase is modeled after BigTable) and LinkedIn uses Voldemort.

- *Apache Oozie*,[16] a Java Web application, supports workflow scheduling operations, helps administrators schedule Hadoop jobs, and is used by Yahoo! Similar products used by Facebook, Google, and LinkedIn include Databee, Evenflow, and Azkaban, respectively.
- *Apache ZooKeeper*[17] is an open-source, centralized service that enables coordination of distributed applications that are stored on servers or with cloud computing. It is used by Yahoo!.
- *Cloudera Hue*[18] is an open-source set of web applications for user queries that interface with HDFS, Hive, Cloudera Impala queries, MapReduce jobs, and Oozie workflows. Hue runs in a web browser and requires no client installation. A comparable product used by Facebook is HiPal. Google uses Dremel.

Copyright note

This chapter contains references to trademarks, registered trademarks, patented solutions and proprietary products. Avro, Flume, HBase, Hive, Kafka, Oozie, Pig, Solr, Sqoop, and Zookeeper are trademarks of The Apache Software Foundation. Hue, Impala, QuickStart, and their logos are trademarks of Cloudera. HiPal and Scribe are trademarks of Facebook. Dremel and Sawzall are trademarks of Google. Azkaban and Voldemort are trademarks of LinkedIn Corporation.

Notes

1 http://lexicon.ft.com/Term?term=big-data, accessed on 19 July 2016.
2 www.forbes.com/sites/gilpress/2014/09/03/12-big-data-definitions-whats-yours/2/#4448536b1fd7, accessed on 19 July 2016.
3 www-01.ibm.com/software/data/bigdata/what-is-big-data.html, accessed on 20 May 2017.
4 https://about.twitter.com/company, accessed on 2 August 2016.
5 www.internetlivestats.com/twitter-statistics/, accessed on 2 August 2016.
6 www.ibmbigdatahub.com/infographic/four-vs-big-data, accessed on 2 August 2016.
7 www.ibmbigdatahub.com/blog/why-only-one-5-vs-big-data-really-matters, accessed on 2 August 2016.
8 www.theguardian.com/money/2013/sep/24/property-websites-alternatives-rightmove, accessed on 20 May 2017.
9 The name Hadoop was originally given to a toy elephant by the 3-year old son of Yahoo developer Doug Cutting in 2006.
10 http://sqoop.apache.org/, accessed on 25 May 2017.
11 https://flume.apache.org/, accessed on 25 May 2017.
12 https://pig.apache.org/, accessed on 25 May 2017.
13 https://hive.apache.org/, accessed on 25 May 2017.
14 https://hbase.apache.org/, accessed on 25 May 2017.
15 http://lucene.apache.org/solr/, accessed on 25 May 2017.
16 http://oozie.apache.org/, accessed on 25 May 2017.
17 https://zookeeper.apache.org/, accessed on 25 May 2017.
18 http://gethue.com/, accessed on 25 May 2017.

References

Boyd, D. & Crawford, K. (2012). Critical questions for big data: Provocations for a cultural, technological, and scholarly phenomenon. *Information, Communication, and Society*, *15*(5), 662–679.

Burkholder, L. (ed.). (1992). *Philosophy and the Computer.* Boulder, San Francisco and Oxford: Westview Press.

Crosby, M., Nachiappan, P., Verma, S., & Kalyanaraman, V. (2015). Blockchain Technology: Beyond Bitcoin. *University of California – Berkeley, Sutardja Center for Entrepreneurship & Technology Technical Paper.* http://scet.berkeley.edu/wp-content/uploads/BlockchainPaper.pdf, accessed on May 20, 2017.

Laney, D. (2001, February 6). 3-D Data Management: Controlling Data Volume, Velocity, and Variety. *Application Delivery Strategies*, META Group.

Lazer, D. & Kennedy, R. (2015, October 10). What We Can Learn from the Epic Failure of Google Flu Trends. *Wired.* www.wired.com/2015/10/can-learn-epic-failure-google-flu-trends/, accessed on August 4, 2016.

Lewis, M. (2014). *Flash Boys: A Wall Street Revolt.* New York, NY: W.W. Norton.

Madden, S. (2012). From databases to big data. *IEEE Internet Computing*, May/June, 4–6.

Sonka, S. (2014). Big data and the Ag sector: More than lots of numbers. *International Food and Agribusiness Management Review*, *17*(1), 1–20.

Ward, J. & Barker, A. (2013). Undefined By Data: A Survey of Big Data Definitions. *arXiv*. 1309.5821.

Part II

Tools and processes

3 Software tools for real estate analysis

One of the primary issues in research, especially in the era of big data, is data management. Without the appropriate tools, the process of collecting, storing, and retrieving information is not only problematic and labor intensive, it has the potential to undermine even the best research. A closely related issue is choosing the right tool for the analysis of the data. There are a number of statistical software packages available, some that are more suitable for beginners with fairly straightforward research projects and others that are sophisticated, powerful, and less intuitive. In this chapter we provide an overview of some of the data management tools and statistical software that a researcher is likely to consider using when they tackle a research project. We base this on the assumption that understanding the tools is secondary only to understanding the research process itself.

Tool choice

When choosing the appropriate data management tool and statistical software for a given research project, there are numerous avenues a researcher can pursue including proprietary, commercial, and open source alternatives. While understanding the options is important, the primary thing to recognize is that the tools are not independent of the research. The research influences what software and other tools should be used to produce the most reliable and reproducible outcomes. In some cases software and statistical programs are of no real importance as long as the experiment is correctly designed and the results properly interpreted. In other cases, the main problem for the researcher is picking the right tool for the job. Furthermore, some data management and statistical research can be completed using only one tool, while in others it may be necessary to incorporate multiple instruments.

With these things in mind and because the options are seemingly endless, the first step for the researcher is to prioritize familiarity with the data and research objectives before settling on data management and statistical tools. Some data are better suited for certain applications than others, thus it behooves the researcher to thoughtfully consider the following questions prior to investing time, effort, and potentially a significant amount of money in ineffective or inappropriate tools.

Research requirements

The first item the researcher needs to consider is the requirements of the research. Not only do the research question and objectives need to be clearly stated, but any challenges or perceived issues also need to be addressed. For example, the type of data, be it textual, numeric, or a blend, can dictate what type of software is used, as can the size of the data set, the frequency of updates, and the consistency of the raw data. Some software applications are more suitable for smaller sets of data, while other applications accommodate larger sets. The requirements of the research can be broken down into a series of specific questions including:

- What is the research question, and what are the primary issues the research seeks to address?
- What type of data is being used? Are they textual, numeric, or a blend/mixture?
- How big is the data set?
- How often are the data being updated? Are they constantly streaming? What is the velocity of the data?
- What other issues are there with the data? Are any missing? Do they need to be truncated? How will inconsistencies be handled?
- What secondary analyses may be necessary to support primary findings?

Reproducibility

The concept of reproducibility has to do with the process associated with analyzing the data. Academic research is typically published with the understanding that any other researcher with similar skills and the same set of data can generate identical findings. When this is not the case, the research can be censored or in more serious cases the findings deemed irrelevant or incorrect.[1] This makes transparency an integral part of the research process and a notable consideration when selecting statistical software. In some cases the use of tools with opaque algorithms is acceptable, while in other cases transparency is of primary importance. Ideally, in the context of reproducibility, the researcher needs to consider the following:

- Is the research process clearly defined and reproducibility important?
- Do the analytics (including the algorithms embedded in a software) need to be transparent?
- Can a researcher with a similar skill set easily replicate the findings using the exact data and processes?
- Will the analysis be re-done or updated in the future?

The context and the skill set of the user

The context of the research, in combination with the skill set of the user, plays an important role in determining what tools should be used. A class project generally requires less rigor than an academic paper, for instance, and professional reports are often prepared using whatever proprietary software a company has licensed, avoiding any semblance of the detail required by academics. Students and less quantitatively minded industry practitioners who are more interested in the bottom line are often (though not always) less experienced with statistical analysis compared to academics and industry-based analysts. Further, the nuances of the various tools make some more suitable for advanced researchers whereas others are intuitive enough for beginners.

Technical support also plays an important role when discussing the skill set of users. With free software, the primary source of support is often a Web-based user group. While this may be sufficient for some, others may require the more formal support that commercial and propriety packages provide. Students may be able to access research faculty for helpful advice while some companies employee full-time staff to handle inquiries. When navigating context and skill sets, the researcher should consider:

- Who is the target audience? Is it for a class project, professional report, or academic journal?
- What type of training does the researcher have? Are there opportunities for additional training? Is technical support available?

Cost

Some tools are obtained at significant cost whereas others are freely available. Commercial software licenses can be several thousand dollars and are limited in duration, but even the most expensive ones do not necessarily have the capacity to support all types of research. Open source is often free, but it may lack necessary technical support, and proprietary packages that may be useful are not always distributed outside of the organization that holds the copyright. When considering value for money, the researcher should ask:

- Does the software (or other tool) support the type of analysis that is proposed?
- Are there any optional items that can be included/excluded (e.g., cloud storage)?
- How long is the duration of the license? Will it cover the proposed research timeframe?
- How frequently is the software updated?
- What type of technical support is available?
- What is the initial cost? What is the cost to renew or extend the license?

The remainder of this chapter is devoted to identifying and briefly describing the statistical software and other data management tools that are most frequently used in real estate research. The list is not exhaustive; rather it is designed to introduce the reader to some of the more widely known options that are applicable to the discipline.

Basic tools

The most basic software tool in the researcher's tool kit is the spreadsheet. Spreadsheets are electronic documents that consist of a collection of rows and columns in a table format. The information stored in the spreadsheet can be numeric, textual, or a mixture of the two, and can be manipulated and/or used for calculations. Spreadsheets are useful organization tools, particularly when data need to be categorized in a particular manner. Recent advances in software have turned the basic spreadsheet into a much more comprehensive tool with a number of useful benefits that vary by product.

Excel

Microsoft Excel is a spreadsheet software designed to interface with Windows, macOS, Android, and iOS operating systems and is most commonly used in real estate research to support datasets. It is not a statistical analysis software per se, but Excel does contain some basic tools that provide simple inferences, with pivot tables, graphing tools, descriptive statistics, and regression models easily created thanks to the user-friendly graphical interface. Recent versions interface with other programming languages and many useful add-ins are available to manipulate data. Of all the data manipulation software available, Excel is probably the most recognizable and is suitable for researchers with little knowledge or background in statistical analysis and/or programming. As long as Excel calculations meet the analytical and statistical needs of the research, it can be very useful.

It should be noted that while it is relatively easy to use to create basic statistical information, particularly for the inexperienced researcher, Excel possesses a number of limitations that make it less desirable when conducting more advanced or complicated analysis. Even though it can be used to manipulate data, large, complex data sets may require other more powerful options. Predictive, inferential, and non-probabilistic statistics are difficult to compute, and the Excel algorithms are opaque, making them more difficult to replicate when compared to other programs. This makes it problematic to share processes in a simple, compact manner. Logic is often easier to follow when using statistical software with transparent code such SAS, Stata, Python, or R.

Argus

Argus is one of the most familiar names in commercial real estate, known primarily for software that supports property valuations and (discounted) cash flow analyses. It is commonly used by many large brokerage firms and possesses a series of standard functions operated by predetermined algorithms including lease analyses, budget and forecasting scenarios, and reporting. Its strictly defined inputs make it suitable for modeling property-level cash flows, rent rolls, and sensitivity analyses. Debt is easily modeled when dealing with very simple structures; though very specific attributes/sensitivities of an asset can be more difficult to model.

The downside to Argus is that the algorithms are rigid and opaque; if the research requires models that necessitate that specific algorithm, then there are no issues. However, when modeling something slightly different or modeling a more creative use of capital, for instance, problems may arise requiring the use of an alternative such as Excel, where the formulas are entered manually.

Commercial statistical packages

There are a number of commercial software providers that specialize in statistical analysis. Most of these programs are comprehensive with user-friendly interfaces covering a range of features from basic linear models to more sophisticated, multi-layered analyses. They are not discipline-specific; rather they support research in many areas such as the behavioral sciences, medicine, public policy, and economics. Below we cover a few of the most well-known and relevant packages to real estate analysis, including Stata, SAS, IBM SPSS, and Minitab. We also provide a brief overview of MATLAB, a proprietary programming language, and ArcGIS, best known for its mapping systems. While not geared toward statistical analysis per se, both packages support and/or supplement research in the real estate discipline.

IBM SPSS

IBM SPSS is relatively easy to use and a favorite among beginners. The intuitive point and click interface supports creative, customizable graphs, and all of the basic statistical functions. The available statistical procedures are accessible from its macro language, as well as its menu-driven graphical user interface. While SPSS does permit an unlimited number of variables in each file, data management is somewhat limited as it is difficult to work with and edit multiple files at the same time. Generally speaking, SPSS ranks as the simplest and easiest to use of all the commercial options, but advanced users may find some of its products lack the in-depth statistical procedures provided by some of the others. Acquisition by IBM in the 2000s triggered its evolution towards integrated enterprise systems and its suite of products range from the simpler IBM SPSS Statistics to the more advanced IBM SPSS Modeler and IBM SPSS Analytics Server.

Stata

Stata is a data analysis and statistical software used primarily in the social sciences and biomedicine. It is often favored by investigators with less background in statistical analysis because it is relatively easy to learn and performs most of the basic statistical assessments. Mistakes are easily identified and corrected and high-quality graphs straightforwardly created using commands or a point and click interface. Commands are reasonably intuitive and uncomplicated making Stata a well-respected and powerful tool in the early career research arsenal.[2]

SAS

SAS is generally considered more appropriate for more advanced users due to less intuitive commands and greater difficulty identifying errors. The learning curve for its very comprehensive procedural language is rather steep, but when surmounted, SAS users have a powerful tool that facilitates complex data management procedures, sophisticated graphing, and the ability to work with multiple files simultaneously. Exceptionally large sets of data can be managed and manipulated in virtually any manner, with the number of observations generally limited to the capacity of the hard drive on the computer. The SAS Institute offers a growing number of products in analytics that range from the traditional SAS/STAT to the more contemporary SAS Enterprise Miner, SAS Visual Analytics, SAS Visual Statistics, and the cloud-based SAS Viya platform.

Minitab

Like Stata and IBM SPSS, Minitab is a relatively easy to use and relatively affordable statistical software package. It offers a menu-driven graphical user interface and a standard suite of analytic and statistical procedures, with a strong emphasis on quality control and experimental design. At the same time, it supports customized procedures through its macro language.

MATLAB

MATLAB differs from the other software packages in that it is a proprietary programming language rather than a statistical package. The foundations of MATLAB are in computational mathematics, hence its popularity in engineering and the hard sciences. It is most commonly used in research related to computations and the physical sciences and supports extensive toolkits for image processing. While it can be used in statistical analysis, MATLAB requires a fairly sophisticated knowledge of mathematics and programming, making it more suitable for the advanced user.

ArcGIS

ArcGIS is a proprietary Geographic Information System (GIS) developed by ESRI. ArcGIS is described as a secure and scalable "complete, mapping platform."[3] It is typically used to build maps and to analyze spatially-related data such as demographics and neighborhood characteristics. Maps are separated into layers, with each layer containing individual features (topography, streets, or census tracts, for example) that can be layered to provide a visual representation of select characteristics of a particular area. Spatially referenced data files from spreadsheets can be precisely incorporated into the maps as long as locational coordinates (latitude and longitude) are included as one of the variables. ArcGIS is frequently used in real estate research to identify and analyze the impact of spatial relationships on property prices and the spatial distribution of economic activities.

Open source players

Open source software is software with a source code that has been developed for the purpose of exchange, modification, and distribution. Use of the software is unrestricted; users can create their own customized versions of the software without violating any licensing agreement or other restrictions. The open source applications emerging as those most commonly used in real estate include R, Python, and QGIS. Closed source software, conversely, is software subject to licensing agreements, with most Microsoft or ESRI products qualifying as the best known in real estate circles.

While open source applications are often described as "free," that does not mean there is no expectation of financial compensation. "Free," in this case, refers to "freedom" – the freedom to do what the user likes with the software. Theoretically, users can charge for any customized version that is created. Note that "freeware" is not the equivalent of open source. Freeware generally refers to software that is available at no charge.

R

The foundations of R lie in the S programming language, one of many products developed by Bell Laboratories (now Lucent Technologies). R provides an open source alternative to S, with both commonly used for a variety of statistical applications such as linear and nonlinear modeling, classical statistical tests, and time-series analyses. Because it is extensible, R can be modified to accommodate a range of research questions; developers can easily add new capabilities allowing it to be customized to their application. This often makes it the tool of choice when transparency and reproducibility are key considerations of the research.

Source code for R is available under the terms of the Free Software Foundation's GNU General Public License. It runs on a number of commonly used platforms including UNIX, Windows, and MacOS. Additional information about R can be found on the website www.r-project.org.[4]

Python

Python is an "interpreted, object-oriented, high-level" programming language with syntax that is easy to use and understand.[5] Python also contains an internal debugger, making its edit-test-debug cycle quick, efficient, and helpful to the novice user. Errors are noted with an exception, and if the exception fails, a stack trace is printed. Like R and other open source players, Python can be freely distributed and modified to meet the needs of the user. The Python source code is available at no charge via the Python Software Foundation at www.python.org. Like R, user-contributed packages extend the possible uses of the software so that it can be modified to meet the specific needs of a research project.

Scala

Scala[6] is a multi-paradigm programming language that smoothly integrates features from object-oriented and functional languages, built on top of the Java virtual machine (JVM). Scala, which is an acronym for scalable language, was developed in 2001 at the École Polytechnique Fédérale de Lausanne, in Switzerland. It is an object-oriented language because every value is an object and every operation is a method call, with types and behavior of objects described by classes and traits. It is also a functional language, which, unlike traditional functional programming languages, allows the user to start writing code in object-oriented style and gradually progress to functional programming. Many companies including Intel, Twitter, and LinkedIn, have switched some or all of their development activities to Scala.

Spark

Apache Spark[7] is a user-friendly, fast engine for large-scale data analytics. Because it supports parallel processing and in-memory analytics, it can be significantly faster than other analytic platforms. Spark can be accessed interactively from Java, Scala, Python, and R shells. Spark uses operations that follow the "map" and "reduce" concepts in Map-Reduce, but its code is compact and more user-friendly than Map-Reduce. Spark's primary abstraction is a distributed collection of items called a Resilient Distributed Dataset (RDD). RDDs have actions, which return values, and transformations, which return pointers to new RDDs. Its built-in libraries include a module for working with structured data called Spark SQL. This loads and queries data from a variety of sources, including Apache Hive tables, and seamlessly mixes SQL queries with Spark programs. Spark Streaming, another built-in library, makes it easy to build scalable fault-tolerant streaming applications. Spark's Machine Learning library (MLlib) contains high-quality algorithms that leverage iterative computation to deliver very fast execution. Finally, GraphX, Spark's Application Programming Interface (API) for graphs and graph-parallel computation, unifies Extract-Transform-Load (ETL), exploratory analysis, and iterative graph computation in a single system that performs on par with the fastest specialized graph processing systems. As a final comment, it is worth mentioning the R package sparklyr,[8] which provides access to Spark datasets and brings them into the R environment for analysis, visualization, and very efficient execution of powerful distributed machine learning algorithms.

QGIS

Quantum GIS (QGIS) is an open source player in the Geographic Information Systems (GIS) market. Like its commercial counterpart ArcGIS, QGIS allows for the management, display, and analysis of geographic information and supports most geospatial vector and raster files. Maps and demographic information can be combined with GPS data (via a plugin), for example, and then be customized to meet the demands of the end user. QGIS integrates well with other open source GIS software and is available across most major operating systems, unlike many of the other open source mapping options. QGIS is a project of the Open Source Geospatial Foundation and is regularly updated by a group of volunteer developers. Additional information, as well as downloads, can be found at www.qgis.org.[9] QGIS is also extensible due to the contribution of its many user-developers.

Database systems

While there are a number of challenges working with big data, one of the most pronounced is the struggle to capture updates or changes to data that are constantly streaming and being created (the "velocity" of big data). As databases grow, finding specific information becomes increasingly difficult even when the data are highly organized. SQL (Structured Query Language) is the specialized programming language designed to address this issue. Using relational algebra, SQL communicates with relational databases (data that are organized in tables according to relationships between the data) to search for information and perform tasks including regular updates and information retrieval.[10] The American National Standards Institute (ANSI) calls SQL the standard language for relational database management systems. Some common systems that use SQL are: Oracle, MySQL, Microsoft SQL Server, Access, and PostGres. Although most database systems use it and many have added proprietary extensions, the SQL commands "Select," "Insert," "Update," "Delete," "Create," and "Drop" are standard among them.

Database systems are typically divided into two categories, either desktop management tools or server-based systems. Anyone who owns a copy of Microsoft Office Professional already owns Access, a desktop data management tool. Access can be used to organize data in a relatively straightforward way, as it does not require knowledge of any programming language due to its user-friendly interface. Its capabilities are good and it can be used to manage a fairly large amount of data.

Depending on the size of the organization and the type of research being conducted, a desktop tool like Access may provide insufficient capacity and management capability for the data that are being collected and analyzed. If that is the case, a server-based tool should be considered. Server-based data management tools offer a number of benefits as they can be modified for custom applications, are powerful, and can be scaled to meet increasing needs in data collection, particularly data that are constantly being updated. On the downside, many server database management tools like Microsoft SQL Server and Oracle come with a hefty price tag. Budget limitations may require the use of open source options that are available at no cost but lack the technological support of fee-based options.

Access

At first glance, Microsoft Access appears to be nothing but an enhanced version of Excel, a place to house data sets that exceed the size limitations imposed in Excel. In actuality Access is a powerful information management tool that reduces the errors and inconsistencies associated with other spreadsheet applications and offers an excellent starting point for beginners who want to understand relational databases and SQL. Tables and fields can be set up to accept only specific types of data, and fields can be grouped together and/or hidden to minimize data input errors. While the latest version of Excel accepts just over a million rows of data, Access supports an almost unlimited amount of records, up to 2 GB per table. Creating links between different items or concepts is useful when investigating relationships within data and reporting is generally easier than with Excel.

Oracle

Oracle is, in many respects, the oldest of SQL enterprises. Known primarily as a multinational computer corporation, Oracle offers a vast array of technological products and services and is second only to Microsoft Corporation in terms of total revenue generated by software providers. OracleSQL is the structured language query available for commercial applications that is developed and marketed by the company.

MySQL

MySQL is one of the most popular and well known open source database management systems. Managed and operated by the Oracle Corporation, it is freely available at no charge under the GNU General Public License agreement for non-commercialized applications.[11] MySQL is used by many high-profile companies including Facebook, Twitter, YouTube, and Yahoo![12]

Microsoft SQL

Microsoft SQL Server is a dual-layered system that includes a database engine and a management studio. The database engine is an invisible service running in the background of the computer, whereas the management studio is a graphical interface that provides access to the information collected by the database engine. Information can be modified, managed, and administered through the management studio. Microsoft SQL can be installed on a server, client, or both and is available through the Microsoft Corporation.

PostGres

PostGres is an open source relational database management option, freely available under the PostGreSQL license that permits the unlimited use, modification, and distribution of the source code for individual or commercial uses. System development began in 1986 with computer science professor Michael Stonebraker and continued by two graduate students who added SQL extensions to what was then known as POSTQUEL, renaming it PostGres95.[13] By 1996, PostGres had entered the open source realm, where it was adopted by a global group of volunteers who changed and improved it and continue to do so. PostGres is used by many diverse organizations including ADP, CISCO, NOAA, and the U.S. Forest Service.[14]

Research "stack"

For any moderately complex real estate analysis, it is unlikely that any single one of the tools discussed above will solve the entirety of your research needs, from data management through the communication of the results. Realistically, some combination of the above will probably be necessary. Within the rapidly expanding community of quantitative researchers, data scientists, and applied computer programmers, the combination of tools that are used for any given project or product are referred to as a "stack."[15]

Stacks can vary from the combination of two relatively antagonistic products that do not offer easy integration functionality to a complex set of integrated data gathering, management, analysis, and reporting tools. Many of the proprietary products have been working towards vertical integration, where the research functions are brought into the same product. Integration of open source tools has been both vertical (but between products) as well as horizontal, in which products that perform the same functions can seamlessly work together. Researchers in industry or research labs may be limited by the stack that their company or team currently employs; for more independent or academic-oriented projects, software costs may limit choice.

There are many possible configurations of the tools discussed above that can constitute a stack. There is no perfect stack and researchers should begin with tools they know and understand well. As the project gains in complexity or repetition, greater efficiency may be gained by either increasing the specificity of the software products in the stack or by moving to a more vertically integrated model. Efficiency in the research process comes with experience and with adding new, knowledgeable members to the team. Choosing your stack is important, but, following the law of parsimony or Occam's razor, a simpler stack that works is better than a more complex one that requires unnecessary maintenance and overhead.

Conclusion (and a word of caution)

The seemingly endless supply of software and other tools available to researchers has grown exponentially over the past few decades. Providers are faced with some of the same problems as researchers – keeping up with the rapid pace at which the amount, type, and source of data changes. These changes require companies to anticipate user needs and engage in innovation to remain competitive in a market peppered with quality products available at little or no cost. Researchers have benefited from this as the availability and access to the number of products has improved markedly, although the greater selection makes choosing the right product more important and less straightforward than it has ever been.

The most important thing to keep in mind, however, is the old adage "garbage in, garbage out." Even the best software, statistical tool, or data management product cannot compensate for incompetency on the part of the user. Incorrect results may be produced from *any* tool if a researcher exercises poor working habits. Research should not only be reproducible, it should be based on the appropriate assumptions and the best available data. A poor investment or bad assumptions are as bad in Argus as they are in Excel, on paper, or with any other instrument.

Copyright note

This chapter contains references to trademarks, registered trademarks, patented solutions and proprietary products. Excel, Argus, Stata, SAS, IBM SPSS, Minitab, MATLAB, ArcGIS, R, Python, Scala, Spark, and QGIS, Access, SQL, MySQL, Microsoft SQL, Oracle, and PostGres, are trademarks or registered trademarks of their respective companies, institutes, or organisations.

Notes

1 Arguably the most prominent example of this is the work done by Donohue and Levitt (2001), which showed that legalized abortion decreased crime. Then, a working paper by Foote and Goetz (2005) showed the effects of a programming error found in a table in the 2001 article. Finally, the sequence of papers resolved with Donohue and Levitt's (2008) article acknowledging the programming error in the 2001 and providing additional econometric evidence of the abortion-crime relationship using differently constructed variables.
2 www.ats.ucla.edu/stat/mult_pkg/compare_packages.htm, accessed on 25 November 2016.
3 www.arcgis.com/features/index.html, accessed on 25 November 2016.
4 www.r-project.org, accessed on 16 November 2016.
5 www.python.org/doc/essays/blurb/, accessed on 22 November 2016.
6 www.scala-lang.org/, accessed on 25 May 2017.
7 https://spark.apache.org/, accessed on 25 May 2017.
8 https://cran.r-project.org/web/packages/sparklyr/index.html, accessed on 25 May 2017.
9 http://spatialnews.geocomm.com/articles/qgisreview/, accessed on 23 November 2016.
10 Punit, J. V., B. Rajput, & A. Sayankar. 2015. Implications for data analyses and higher education with SQL and data analysis. *International Journal of Engineering Applications and Technology*, April, 80–86.
11 https://dev.mysql.com/doc/refman/5.7/en/what-is-mysql.html, accessed on 1 December 2016.
12 www.mysql.com/about/, accessed on 1 December 2016.
13 www.postgresql.org/about/, accessed on 30 November 2016.
14 www.postgresql.org/about/history/, accessed on 30 November 2016.
15 Our use of the term "stack" must not be confused with other uses of the term, such as Stack Exchange (https://stackexchange.com), which is the network of question and answer (Q&A) communities that answer questions about software development.

References

Donohue, J. & Levitt, S. (2001). The impact of legalized abortion on crime. *Quarterly Journal of Economics*, *116*(2), 379–420.

Donohue, J. & Levitt, S. (2008). Measurement error, legalized abortion, and the decline in crime: A response to Foote and Goetz. *Quarterly Journal of Economics*, *123*(1), 425–440.

Foote, C. & Goetz, C. (2005, November 22). *Testing economic hypotheses with state-level data: A comment on Donohue and Levitt (2001).* Federal Reserve Bank of Boston Working Paper 05–15.

Fox, J. & Leanage, A. (2016). R and the Journal of Statistical Software. *Journal of Statistical Software*, *73*(2), 1–13.

4 The analytical workflow

As data availability has grown, so too has the complexity of real estate analyses that are commonly performed. The typical real estate research or analytical project that was once a simple model using a single data source over a short time period is now a multi-faceted set of longitudinal models that leverage a handful of disparate datasets, geographic information systems (GIS) analyses, and live streams of data. With this expansion in scope the importance of planning an efficient and transparent research process or analytical workflow has never been greater.

In this chapter we will outline the key steps or phases in an efficient workflow for a complex research project. We will focus on common tasks and steps, as it is impossible or improbable to properly fit every research project into our scheme. We offer guidelines that are to be loosely followed but that can also be altered or outright ignored if the analysis at hand calls for it.

After an explanation of the general workflow outline, we focus on two issues of particular importance when organizing the research process: (1) the how, when, and why of removing observations from the data (often referred to as data cleaning); and (2) the rationale for conducting real estate research and analysis in a reproducible manner.

Workflow outline

We see most moderately complex analytical workflows in the real estate field to have six major phases or steps; Gather, Manage, Prepare, Explore, Clean, and Analyze.[1] Those who are inexperienced with complex analytical projects often envision the project as consisting of 10% data gathering and 90% data analysis, with little or nothing in between. In reality, we have found that a considerable amount of time and resources are spent on what happens between these two steps.

One possible explanation for the confusion surrounding the set of tasks that occur between data gathering and data analysis is the lack of a shared language or set of terms to describe this part of the research process. Several comprehensive terms have been proposed to describe these tasks – 'wrangling,' 'munging,' 'curation,' 'manipulation,' 'tidying,' 'grappling,' and 'cleaning' – all with their shortcomings (Horton, Baumer, & Wickham, 2015). The term 'data munging' has gained acceptance throughout much of the applied data science field (Cady, 2017; Gutierrez, 2015), so we will use that term to describe these intermediary tasks. Figure 4.1 shows the oft-imagined workflow alongside what we believe to be a more honest description of a comprehensive analysis, complete with a series of data munging tasks.

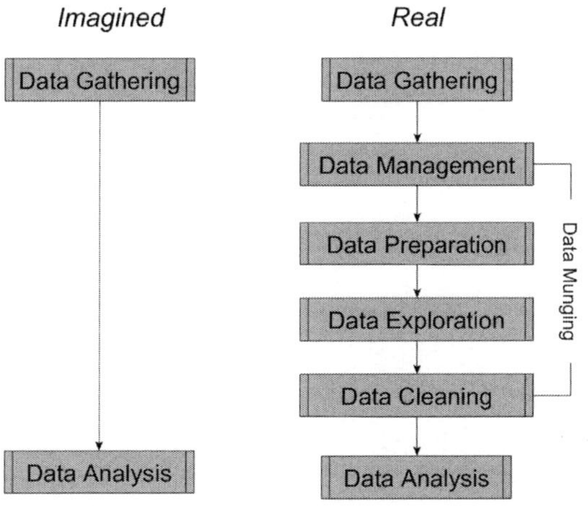

Figure 4.1 The analytical workflow: imagined and real

Chapters 5 through 11 of this book present an in-depth discussion of these six steps:

- Gather (Chapter 5): Obtaining the raw data and licensing for the necessary inputs to the project
- Manage (Chapter 6): Properly compiling and storing the data for maximum efficiency, transparency, and reuse
- Prepare (Chapter 7): Getting the data ready for investigation
- Explore (Chapter 8): Familiarizing yourself with the data, identifying potential issues
- Clean (Chapter 9): The removal or correction of remaining discordant values (data errors and outliers)
- Analyze (Chapters 10 and 11): Complex analysis of numerical and textual data in real estate

A visual representation of the proposed analytical workflow can be an invaluable part of the research design or project scoping process. Rarely is it correct the first time around, and as the project progresses the diagram of the process should be adjusted to keep pace. Throughout the remainder of the book, when we refer to 'analytical workflow' it is this sequence of six processes – gather, manage, prepare, explore, clean, analyze – to which we are referring.

Removing observations from the data

With the possible exception of choosing a research design, the removal of errors and discordant values from the data is one of the most critical steps in the research process (Chapman, 2005). The extent to which the data are 'cleaned' prior to analysis can have as large of an impact (sometimes larger) on the final results as the choice of statistical test, model specification, or diagnostics. Despite this, the process of removing observations is one of the least discussed and documented components of quantitative research design (Krause & Lipscomb, 2016). With the rise of big data and data-driven methods of scientific inquiry

Table 4.1 Observation removal rationale

Rationale	*When to Remove*	*Objectivity*
Error of Structure	Data Management	Objective
Relevancy	Data Filtering	Objective
Error of Join Matching	Data Filtering	Objective
Error of Observational Inconsistency	Data Filtering	Mostly Objective
Error of Value and Missing Data	EDA/Data Cleaning	Mostly Objective
Discordant Values (Outliers)	Data Cleaning	Somewhat Subjective

(Gray, 2009; Nelson, 2009), the criticality of properly handling data issues and errors is only growing.

It can be tempting in the data management phase to undertake the process of removing observations from the data, but this is not advisable due to reasons of efficiency, validity, and the future re-usability of the data. In our view, the data are continuously improved from raw to ready to analyze by following the workflow shown in Figure 4.1.

The term 'data cleaning' is often presented as a catch-all term for removing observations from a data set. As such, data cleaning can carry with it a negative connotation, whereby the observations are systematically eliminated until a desired result is achieved, but this approach has the potential to seriously undermine the validity of the research. Rather than approach data cleaning from a results-oriented perspective, the process should be conducted in a staged approach and be seen as a sequence of sub-processes guided by different principles. This process should follow some standard and objective guidelines to the extent that is possible. Doing so can help legitimize the analysis as well as facilitate greater collaboration.

Table 4.1 presents six rationales for removing an observation from a dataset, listed in general order of occurrence, though the exact order of the operations may vary based on the idiosyncrasies of the project. Also listed is a measure of objectivity for each rationale that is intended to gauge the overall defensibility of removals at each step. Those with more subjective rationale may need to be validated with sensitivity tests (discussed more fully in Chapter 10). Also note that not all errors automatically require removal of the observations. In cases where the errors can be corrected with efficiency and certainty, the correction should take place.

- **Error of structure.** Errors of structure refer to cases where a value(s) for an observation is of an incorrect data type, does not fit with the data types of the other values within that field, and/or is incongruent with the data management scheme. An example would be an observation that had a value of 'large' in a numeric field that represents home size in square feet. While mistakes should be identified and corrections made, it is not always possible to accurately do so, thus the offending observation may require removal. This is an objective form of data removal. If the error of structure is present in the raw data, then the removal of the offending observation(s) occurs in the data management phase. If errors of structure arise after appending data from different sources, corrections or data removals will occur in the data preparation phase.
- **Relevancy.** Removal of observations due to relevancy involves cases where an observation is outside of the scope of the research question or analysis. For example,

if a study is focused on office rents and the data include observations of retail sales, the retail sales observations can be removed for reasons of relevancy. Removal due to relevancy is done at the data preparation stage and can be considered an objective removal of observations.

- **Error of join matching.** Errors of join matching occur when data integration results in systematic and total missing-ness of a particular field conditional on the joined field. In other words, if two tables are joined and the values of the unique identifiers are not equivalent across the two tables then the resulting joined data will contain systematic missing-ness in one or more field due to the joining process. Incomplete joins of this type do not always represent an error, especially in cases where the target table is smaller in scope (within the unique identifying field) than the new (to be joined) table. An example of a matching-induced error would be a case where crime rates for a given neighborhood are joined to housing sales transactions. If information on crime rates is missing for a given neighborhood then the join will create missing data for sales transactions in those missing neighborhoods. Removal due to join matching issues occurs in the data preparation phase and can be considered mostly objective.
- **Error of duplication or inconsistency.** When appending data from different data sources, duplication or inconsistencies of the same observation can result, e.g., there may be two observations (one from each source) relating to the same building. In the case of duplicates, one of the observations can be removed. In the case of inconsistencies, the observations refer to the same entity but individual values differ. For example, one data source states a building is 1,400 square feet and another data source describes it as 1,300 square feet. Obviously both cannot be correct so one of the observations must be removed. Choosing the correct observation to remove is discussed in Chapter 7, "Data Preparation." While the removal of duplicates is wholly objective, the removal of inconsistencies may contain some measure of subjectivity.
- **Error of value or missing-ness**. Errors of value occur when an observation contains an individual value that is incorrect to such a degree that either the measurement itself or the scale of measurement is wrong or a key-punch error has occurred. An observation of a single-family home that lists 25 bathrooms is an example of an error of value. Without access to the original data collector (as is usually the case), it is impossible to know if this is actually correct (as it is for 1 Buggy Whip Drive in Rolling Hills, California, USA), an issue of measurement (recorded as total water fixtures – as some jurisdictions measure), or a key-punch error. Either way, this value is of an order of magnitude so different from the remainder of the field that its removal is warranted. Since no value is a wrong value, missing data can also be considered an error of value. Special techniques and metrics are often used to address these issues. Removal for reasons of value generally occur in the Exploratory Data Analysis (EDA) or the Data Cleaning phase, depending on the severity of the value and the type of EDA employed. Removing errors of value can be highly objective, though as the magnitude of difference between the suspect value and the remaining values in the field narrows, objectivity decreases.
- **Discordant values (outliers).** Managing discordant values or outliers is the final and most subjective of the observation removal process. Outliers are correct values occurring on relevant observations but have a value that is discordant or incongruent with the remaining values in the field. A discussion of outlier identification and

handling techniques can be found in Chapter 9. In general, the removal of outliers should be done with care and be thoroughly explained and justified, as the process, even though it usually involves standard statistical tests, is considered subjective. Removal of outlying or discordant observations is typically done in the Data Cleaning phase.

The rationale for removing observations from a dataset will almost always fall into one of the categories listed in Table 4.1. When the reasons are subjective, the rationale should be fully explained and, if possible, assessed for impact on the final outcome with sensitivity testing. Doing so will limit concerns about validity and the intentions of the analyst or researcher to influence the results of the study.

Reproducibility

While the issue of reproducibility (or the lack thereof) has risen out of academia, the need for a transparent and repeatable workflow is a key component of good industry practice as well. A reproducible workflow is one in which the process of conducting an analysis of data (an analytical workflow) is done in such a way that it can be exactly recreated by another researcher, as well as by the original researcher. This means that the research is conducted so methodically that at the end of the analysis there is an instruction manual that will explain exactly how the research was produced, every step of the way. In the ideal case the reproduction of the work would be done in a very automated fashion, such as copying a number of files from one location to another and then hitting a 'submit' button to 're-run' the analysis. By virtue of being repeatable and by having an instruction manual, the workflow will be transparent to others.

A key distinction in describing a workflow is how the individual tasks are completed. Most operations can be categorized as either 'code-based' or 'click-based' (e.g., the difference between using R or Excel for a project, respectively). Code-based operations are those in which a specific, clear, and unambiguous instruction is given to a computer in a language that it can understand. A record of this operation is necessarily stored in the input file or the log file of the computer. Conversely, a click-based operation is one in which a human manually makes a change with a 'point and click' operation, usually with a graphical user interface (GUI)–based software. Point and click operations are generally not saved by the computer and are not fully reversible, regardless of the time between initial operation and required reversion. Having an undo button in a software product is *not* equivalent to a fully reversible set of operations because there is no clear log of operations and because an undo action only lasts for the current session. Additionally, point and click operations generally do not offer 'executability,' or the ability for the entire set of previous operations to be re-done with the click of a button. Executability is not a requirement for reproducibility, but it is a very important step to taking full advantage of the number of the benefits a reproducible workflow provides.

Reproducibility has implications for analyses done in both academia and in industry. From an academic standpoint, reproducibility allows a reviewer or other researcher to validate the results. While this should seem to be an obvious requirement for published work, the vast majority of research that has been published over the past century has not been done in a reproducible manner. Only recently has a movement arisen in academia pushing for greater transparency and reproducibility underlying published studies.

For industry, reproducible output will aid in the daily operations of the firm. Managers can easily verify and validate employee work, and new employees are more easily trained on how particular outputs are generated. In a time of high employee turnover, having a transparent and repeatable workflow can provide a company with a marked financial and human resource advantage. Additionally, if a firm is involved in litigation of any kind, a fully documented and reproducible workflow can be incredibly helpful in legal situations in which past work must be recreated or demonstrated for the court.

The 'why' of reproducibility

There are three main reasons for adopting reproducible practices in real estate-related data analyses:

1 Efficiency
2 Collaboration
3 Expertise

All three reasons apply to both research teams and projects that are academic or industry-focused.

Efficiency

Efficiency is the first and most obvious rationale for implementing a reproducible analytical workflow in real estate. One of the benefits of a reproducible workflow is that it is focused on the process, and the arrival of new or updated data does not necessitate a new process, only a new execution of an existing process. This is especially important in real estate for two reasons. First, most real estate-related analyses or metrics are repeated over time. Monthly home price indexes, quarterly rent growth figures, or bi-annual supply changes are just some examples of real estate metrics or analyses that are repeated multiple times throughout the year. With a reproducible workflow, creating these periodic reports involves only updating the data, not the process itself. Yet, in situations where methodological improvements are made (e.g., price index methodological improvements), efficiency implies that the results of the 'old' methodology can be compared very easily to the 'new' methodology using the same archived data.

Nearly all real estate data has "velocity," which means that new observations are constantly and consistently being created in the market. As in most cases, the current conditions are the most important (and valuable to potential and existing customers), and creating a process that can seamlessly integrate the newest data will save considerable time and money and allow for reporting to be done on the most up-to-date market conditions. Having a process that can handle velocity also allows for the creation of custom market reports that occur outside of the regular periodic publications. In short, there are both cost savings and new revenue potential available with a reproducible workflow.

The efficiency benefits of reproducibility are not limited to industry; they also flow to academic researchers as well. A reproducible workflow makes responding to peer-reviewed comments much easier. With a reproducible workflow there is little difficulty in changing a data filter or method in an early step of a multi-step analysis. The results can then be simply re-run and the manuscript updated. Additionally, most academic projects start and stop multiple times over the course of the project as grant funding lapses, peer review wait times stretch out, and obligations such as teaching and administrative duties disrupt research agendas. Returning to a project from a long hiatus can be made much easier and quicker when the initial work was done in a well-documented and reproducible fashion. Reproducibility has an additional benefit to those who run real estate research centers, too – specifically it facilitates the periodic update of a regional research report on housing market trends as often required as part of a company's sponsorship of a real estate center on university campuses.

Collaboration

Increasingly, both industry and academic research is accomplished by teams, many times with various members working from remote locations. A reproducible workflow can greatly aid in facilitating the collaboration necessary to successfully complete team-based projects. The collaborative benefits of reproducibility are derived from three components – integration, validation, and education.

Reproducible workflows are process-oriented and in most cases produced using code-based software, or at the very least, a detailed set of past activities. Code-based processes that are worked on concurrently by multiple researchers can be easily integrated using a version control system such as Git or Mercurial. These version control systems can also be extended to include collaboration on output such as reports, peer-reviewed articles, and websites. Regardless of the actual systems or specifics of the reproducible workflow used, the mere existence of a complete set of all data operations finished to date allows all team members to clearly identify what has and what has not been done, ultimately leading to enhanced integration of each researchers' contribution to the project.

Teams working in a fully reproducible environment also benefit from easier and more frequent validation of their efforts. Other team members or supervisors can more efficiently check any work to ensure that it is done properly and that it fits the goals and guidelines of the overall project. Over the course of the project, this increased validation will lead to more accurate results. In many cases, a reproducible workflow may also lead to increased creativity within the team as each team member will be better able to envision the whole of the project and add to it with their own specific skills and knowledge.

Finally, reproducible analyses offer an excellent opportunity for the training of new employees or post-doctoral researchers. Successfully completed projects can be used as educational tools for the new researchers in which they can clearly track an example of the entire workflow that is used within the company or team. Asking a new collaborator to reproduce an existing analysis can both introduce them to the specific processes used by the team as well as efficiently test their aptitude in such matters. Another related benefit is that all team members are able to access, visualize, and experiment with the entire workflow process, not just those tasks where they specialize. This process will lead to the inherent growth of employee and researcher skill sets over time.

Expertise

As the size and complexity of datasets increase, so does the need for code-based processes rather than click-based processes. Most big data analytics require specialized software, the majority of which are code-based products. As a firm or team's research expands in scale, a reproducible process becomes progressively more necessary due to the requirements of the software and other tools that are used. Additionally, large-scale analyses often require bringing software engineers, database architects, and/or data scientists into the team. Researchers with these skills are usually well-trained in reproducibility and will only work with tools that allow it. In short, if a company or team is looking to expand its expertise into projects, analyses, and lines of business that are large scale, reproducibility is something that cannot be ignored.

Working reproducibly requires a process-based approach and with this comes process-based thinking. In doing so, tasks can be separated from strategy, and higher-level contributors brought into the process. From a business standpoint, this means better engagement with upper management and other departments in the firm. For academics, senior researchers may be able to better interact with junior contributors in a process-based framework. In either case, the expertise of a wider set of contributors can be leveraged in a project.

Overall, transitioning to a reproducible workflow will likely improve the efficiency, collaboration, and expertise of a team or individual project. The benefits flow to both academic research groups as well as industry-based groups. Many big data analytics require software and personnel that only work this way.

The 'how' of reproducibility

Two key concepts make up the core of a reproducible analysis: data provenance and analytical workflow (Goodman et al., 2014; Krause, 2016). Data provenance is defined as a record of all processes, people, and documents that were involved in the research outcome; it is the instruction manual necessary to recreate the analysis from beginning to end. The analytical workflow is the method of using the data provenance, in its own particular order, to create the output from the input. Data provenance is a thing while the analytical workflow is a process.

Data provenance involves the faithful recording of every step, from data collection and cleaning to the production of figures and tables. It should perfectly and unerringly connect the raw data to the final output, including all data cleaning, preliminary models, model output, diagnostic and sensitivity tests, figures, tables, and any narrative (sometimes called 'comments' embedded in the source code) that may be necessary to properly understand what was done. There is no single form that a data provenance must take. In practice, the specifics of the data provenance will depend heavily on the type of analysis done, the output, the data, and the software used in the analysis.

The analytical workflow is the process followed to create the final results. Like the data provenance, the specifics of the analytical workflow depend on the analysis undertaken. In terms of efficiency, the most useful type of analytical workflow is one that is executable. Executable means that the entire workflow, from data collection through the reproduction of the final output, can be recreated with the push of a button, or more likely, the running of a script or small program. Often some portion of the workflow, such as data collection, cannot be automated and thus many analytical workflows are semi-executable at best. The greater the executability of a workflow, the easier it is to update an analysis when there is a desire to change the data, methods, output, or other step in the data provenance.

In practice, executability is usually done with code-based processes in which all of the operations are saved in a script. A script is an executable file that runs in a given software language and will accomplish some or all of an analytical workflow. Often a script will work with a given set of data, but it is not flexible enough to handle new or different data sets. More advanced researchers may find themselves writing code that contains more flexible functions that are able to perform the same type of small task over a number of different datasets with a variety of user-defined parameters.

A third term, compendium, has arisen to signify the totality of a research project. A compendium is the entire file folder containing everything about the project, including raw data, data provenance, an executable workflow, and all final outputs. A reproducible research compendium is one that can be transferred to another person or team and exactly replicated. In other words, projects should be designed such that at the end a compendium can be handed over to a non-member, who has everything necessary to replicate and understand the research project. This means that code and scripts need to documented, meta-data faithfully explained, figures and tables annotated, data sources given, and all subjective decisions clearly explained.

Basic skills

Many researchers are already doing some portion of their research in a reproducible manner, so moving to a fully reproducible approach will not require a complete change of habits. Often the addition of a few new skills or processes will be enough. Some of these skills are described below.

First, as mentioned above, code-based operations are much easier to record in a data provenance and, in most cases, necessary to create a fully or semi-executable analytical workflow. Simply put, the first step to reproducibility is learning to code. Reproducible research does not require the most efficient, most complex, or most elegant code. Any code that is understandable by another researcher is good enough. Moving from 'click-based' to 'code-based' research does involve an investment of time, but that time will be quickly regained once the efficiencies of code-based work begin to pay dividends.

A second skill to master is that of improved data storage procedures. Most big data analyses require specialized storage software or systems. While these may be helpful, they are not necessary for reproducibility. What is necessary is that the data be fully documented in terms of the sources, names, and meanings of the fields and other necessary meta-data. Getting in the habit of recording this information will not only make research more reproducible, but it will also help better understand the strengths and weaknesses of the data. Table 4.2 presents an example of a data dictionary – a core component of good data documentation.

Another useful habit to enhance reproducibility is to eliminate the manual altering of data, models, tables, or results. What this means is that all data cleaning, field transformation, model parameter setting, and figure plotting should be done with machine readable rules, not subjective decision-making. Moving to a code-based process will help achieve this goal. In addition to increasing the reproducibility of the work, it forces researchers to

Table 4.2 Example data dictionary

Field Name	*Data Type*	*Description*	*Source*
PIN	Numeric	Unique ID #	County Assessor
TransDate	Date	Date of Sale	County Assessor
TransPrice	Numeric	Sale Price	County Assessor
RecordingNbr	Numeric	Recording ID #	County Assessor
BuyerName	Text	Buyer Name	County Assessor
SellerName	Text	Name of the Seller	County Assessor
PropertyType	Factor	Property Category	County Assessor

think hard about the rationale for those decisions – such as removal of discordant values like outliers or data errors – which will ultimately aid in the defensibility of the research after publication or delivery to a client.

Conclusion

Understanding the extent of the analytical workflow is a critical task at the onset of any large real estate research project. This chapter outlines the six sequential steps that are generally required in a large-scale and/or complex analysis project, providing an in-depth look at producing a reproducible workflow. Detailed discussions for each of the steps are provided in the chapters that follow. The reader should view this chapter as a road map for the remainder of the book.

Reproducible research is important for both academic and industry-based research. Regardless of the sphere of the research, working reproducibly helps to increase efficiency, foster collaboration, and leverage expertise. In short, it saves time and money while creating a more accurate and defendable result.

For most researchers, working more reproducibly involves small changes to existing habits. Moving from a 'click-based' to a 'code-based' process will help to record data provenance and to move closer to a fully executable workflow. Recording meta-data and eliminating subjective analytical decisions (such as outlier removal) will assist in creating a research compendium that is fully transparent, transferable, and defendable. Like all investments, some costs in time, software, and training may need to be undertaken, but they will likely pay for themselves in the long run.

It is not our intention to suggest that the entire research compendium should be automated, and that there is little room left for the human researcher. Quite to the contrary, researcher expertise is still required in the formulation of the research question, the choice of models, the degree of data cleaning, and the manner of visualization of the results. The process of reproducibility merely seeks to automate the annoying steps of re-cleaning, re-modeling, and re-plotting the data every time a small change occurs in an earlier step in the workflow. Finally, regardless of the level of automation in an analysis, the researcher still plays a vital role in interpreting the results and communicating them to the public. Ideally, a reproducible workflow will reduce the amount of time spent on the workflow itself and allow for greater focus on interpretation and communication.

Note

1 One could argue that there are two more that would bookend this list, Design and Communicate. We agree that the initial design of the research question or project and the final communication of the results are incredibly important steps; however, these are outside of the bounds of this book as we are focused on the more mechanical issues of getting from raw data to results.

References

Cady, F. (2017). Data munging: String manipulation, regular expressions, and data cleaning. In *The Data Science Handbook* (pp. 47–60). Hoboken, NJ: John Wiley & Sons.

Chapman, A. D. (2005). *Principles and methods of data cleaning – primary species and species-occurrence data.* Version 1.0. Report for the Global Biodiversity Information Facility, Copenhagen, Denmark.

Goodman, A., Pepe, A., Blocker, A. W., Borgman, C. L., Cranmer, K., & Crosas, M. (2014). Ten simple rules for the care and feeding of scientific data. *PLoS Computational Biology*, *10*(4), e1003542. https://doi.org/10.1371/journal.pcbi.1003542.

Gray, A. J. G. (2009). eScience: A transformed scientific method. In A. J G. Hey, S. Tansley, & K. M. Tolle (Eds.), *The Fourth Paradigm*. (pp. xvii–xxxi), Redmond, WA: Microsoft Research Redmond.

Gutierrez, D. D. (2015). *Machine Learning and Data Science: An Introduction to Statistical Learning Methods with R.* Basking Ridge, NJ: Technics Publications.

Horton, N. J., Baumer, B. S., & Wickham, H. (2015). Setting the Stage for Data Science: Integration of Data Management Skills in Introductory and Second Courses in Statistics. *arXiv Preprint arXiv*. 1502.00318.

Krause, A. (2016). Reproducible research in real estate: A review and an example. *Journal of Real Estate Practice and Education*, *19*(1), 69–85.

Krause, A. & Lipscomb, C. A. (2016). The data preparation process in real estate: Guidance and review. *Journal of Real Estate Practice and Education*, *19*(1), 15–42.

Nelson, M. L. (2009). Data-driven science: A new paradigm? *Educause Review*, *44*(4), 6–12.

5 Gathering real estate data

Gathering data for real estate analyses once consisted of begging local agents for market information, hand copying government records, or waiting for data CDs from a subscription-based service to arrive by mail. The skills necessary to succeed at acquiring data today are decidedly more technical. That is not to say that maintaining good relationships with local market players is not a valuable way to access information, but the process has certainly modernized. Today, the process of data gathering consists of three primary steps: developing a data collection plan, licensing and use assessment, and data procurement.

Data collection plan

Any moderately large analytical project should develop a data collection plan in the early stages of the project. As such, a data collection plan is one of the initial components of the analytical workflow. Like all components of the research process, the plan may be subject to change during the course of the project and should be updated as necessary.

The key rationale for developing a well-defined data collection plan is efficiency. For teams, duplication of data collection is a wasteful exercise that can be avoided with adherence to a collection plan. Likewise, considerable collection of unnecessary data may negatively impact the timing or the financial budget of a project. In cases where data has a limited temporal availability or is replaced periodically, a data collection plan will allow the researcher and the researcher's team to access and procure data at the necessary times. Finally, data procurement can occasionally be a very lengthy process, especially when custom licensing agreements need to be negotiated before data is available. In these situations it is imperative to budget enough time and money to see the acquisition process through to the end.

By outlining a plan, various subsequent activities – data management, preparation, exploration, cleaning, and analysis – may also proceed in parallel with some of the data collection. This is especially important if data collection of some subset of the necessary information is envisioned to be very time consuming or involves data with high volume. As work proceeds to the more advanced steps, any additional data needs can be identified and the necessary data gathered sooner rather than later.

A structured data collection plan should include hypothesized uses for each data set that is deemed necessary for the research. A comprehensive understanding of future data use will help determine the specificity, areal, and temporal extent as well as level of observation needed for each data set. While it is generally best to collect data at its most disaggregated level, if only aggregates are needed and disaggregated data are expensive, extra costs may be avoided by gathering only the aggregated data set.

The majority of large analytical projects in real estate utilize secondary data (i.e., data that are collected by another party or parties). However, if primary data such as interviews, surveys, or on-site observations (such as measuring a building) are necessary, then a data collection plan is imperative. Prior to beginning data collection, thought should be given to what information is necessary to be collected. Special attention should be paid to how values regarding time and space (location) will be measured and recorded. Also, if the collected data are to be integrated with other secondary sources, the collection process must gather information that allows for this integration, such as an address or location (latitude and longitude) for matching purposes. Researchers operating in an academic environment who are collecting data from people (interviews and surveys) may also have human ethics review processes (through entities such as institutional review boards or IRBs) completed prior to beginning the data collection. As compliance with these regulations may be time consuming, a data collection plan will help to build in enough time so that the project deadlines are not missed.

In sum, a data collection plan does not have to be an onerous process. Simply charting the required data sets, the time needed, and dependencies (external and internal) provides the foresight necessary to avoid potentially simple, yet damaging, mistakes in the data collection process. When working in large and/or distributed teams, the data collection plan can serve as a useful guide to various activities that are being conducted simultaneously.

Licensing and use assessment

After the data collection plan has been developed, the licensing and determination of the extent of potential use for any data needs to be properly assessed. In many cases, the allowed uses will be spelled out clearly in the license or purchase contract. While the standard legal agreements are often written for business-to-business transactions, situations involving academic or not-for-profit institutions may be less clear. For example, some academic research funders require the sharing of data along with results at the completion of a project. This requirement can be difficult or impossible to comply with when data are proprietarily licensed. Additionally, sharing data among research collaborators or re-using data for extensions to the original work or for other related projects may prove difficult if the original license was very restrictive on permitted uses.

The purpose to which the data is being put to use can also impact the licensing and the cost of acquisition. Many large data vendors and/or aggregators are willing to provide data at a greatly reduced cost – occasionally free – to academics or non-profit organizations engaging in public (non-commercial) research. As many universities are now actively promoting incubator labs and the creation of commercial products from initial academic research, conflicts of data use are likely.

Even data that are openly available on the Internet, such as data from websites like Zillow.com and Domain.com.au, have use restrictions that users implicitly agree to through use of the site. The details of legal issues surrounding these sources are covered more deeply in Chapter 13; researchers need to fully explore the data source or website fully before assuming free and unfettered use.

Generally speaking, much of what has been traditional real estate data – sales and lease transaction and building characteristics – is often heavily licensed and moderately expensive. For much of the real estate industry, information asymmetries are a key component of the business model; as a result, data are only shared at a price or in a limited manner. Intentional data, such as census information and standard public infrastructure shapefile data, are usually available free and without restrictions on use. The new (big) data sources that are being used may be freely available, but they often come with use restrictions, as discussed.

When developing the analytical workflow and determining data needs, issues regarding licensing and use restrictions should always be considered. In many cases, the licensing process can be slow, and time should be built in for this contingency. Additionally, if researchers are hoping to expand or extend the use of the data to additional projects, these considerations should be addressed at the initial discussions and negotiations regarding data licensing and use.

Procurement

Once the licensing and use limitations have been negotiated and confirmed, the data can be procured. Actually procuring real estate data was once a physical act; now it is an electronic one. In fact, in many cases the researcher many not even directly obtain the data itself (in electronic form) but rather may be allotted access to an existing dataset that is stored on a remote server. To determine a preferred method of data procurement (or access thereto), there are three important considerations: (1) storage requirements; (2) security issues; and (3) repeated access.

For very small datasets, storage on a local computer or server may be of no consequence, and full electronic procurement of the data is the obvious choice. Conversely, for very large datasets or for ones that change rapidly, researchers may wish to obtain access only to the data and not actually store the data onsite. Many cloud-based services now offer both data storage and computational operations such that datasets of many terabytes can be stored off-site and analyzed in the same off-site location. This option can remove the need for expensive hardware purchases and the maintenance required to keep that hardware in working order.

A related and common type of arrangement is one in which a data provider offers access to a database via a searchable front-end website. Many of the multiple listing services (MLS) in the U.S. offer this form of service, whereby appraisers (valuers) and agents subscribe to a service where they can search and download specific subsets of the most recent data but do not have access to all data at one time. Users who do many small analytical projects, often with different data parameters, can benefit from this model of data storage by not being burdened with storing and maintaining a large database and having full, searchable access to the subset of the data that may be required at a given time. One downfall to both of the "access" types of arrangements is that in the event of a network outage, data may be unavailable. This may be particularly troublesome for analysts working while traveling or those who spend time in places where Internet access is slow or expensive.

In most cases, on-site storage of the data is likely to be the preferred option. In these cases, data transfer becomes an important component, especially if there are frequent updates of large datasets. Most large data vendors will offer secure, cloud-based downloading; however, smaller players in the market or data sources for whom data transmission is not their primary business still insist on sending data via email for smaller datasets or on DVDs, flash drives, or through file transfer protocol (FTP) sites for larger datasets.

If the data are sensitive, security can be a concern in both the on-site storage and the access-only situation. In an access-only case, the security risk is via a compromised log-in. On-site, the risk for outright theft is much larger. Though much of real estate data is not highly sensitive, some data, such as financial records, mortgage information, and owner addresses may be sensitive, and the use of property data security measures should not be overlooked. Additionally, many countries have specialized procedures that apply to data that includes personally identifiable information (PII), such as owner names, personal identification numbers, addresses, etc. Data that may include PII should be identified in the data collection plan so that any additional security processes that are necessary can be included in the project budget.

Finally, a major consideration in the data procurement process is whether or not repeated access to the data is required. If an analysis is retrospective and all of the necessary data can be collected at once, then a single procurement exercise may be sufficient. If this is the case, a relatively inefficient method such as a mailed DVD or a collection of emails with subsets of the data may not be overly painful. On the other hand, if the analysis is intended to be contemporary and continuously updated over time, an efficient data procurement method should be a priority. Direct download access via FTP or application programming interface (API) is preferred in these situations so that the procurement process can be automated. Understanding the repetitiousness of data collection should be determined in the data collection plan in the hope that a suitable and efficient process can be agreed upon with the data sources.

As many big data sources such as Twitter and Google move toward APIs, the use of such tools is becoming more important for real estate researchers. An API is a structure or framework that allows different software programs to clearly speak to each other. In the realm of data collection this means allowing the researcher's computer to talk directly to the database held by the data source. In other words, researchers are able to write code to directly query information from another party's database and then, often, download that data directly to their own hard drive or server. For analytical workflows that requires repeated data collection over time, APIs are the most efficient manner of automatically collecting data at pre-specified time periods.

APIs are more useful than an FTP – which is a static website link containing access to a single dataset – because queries can be performed prior to download. What this means is that one does not have to procure the entire dataset (as is the case with FTPs) when only a portion is needed. APIs have become particularly useful because many big data providers like Google and Twitter are not willing to hand over their entire dataset, nor could most users adequately store the data. The API, however, allows these data owners to allow access (public or paid) to their data whereby the total data access can be limited, and the process of specifying queries and transferring data lies with the potential user and the not the data source.

Web scraping

In the absence of an available FTP or API, some data that are displayed on the web can be retrieved by reading the raw HTML files displayed in the web browser. This systematic gathering of data from the Internet is referred to as "web scraping." Here we present the practicalities of web scraping, leaving the researcher to determine if this is an appropriate technique for their work. The legalities (or ethical considerations) of the practice are discussed in Chapter 13. For many websites, scraping or electronically harvesting data is addressed in the site's terms and conditions.

Again, web scraping works by reading the raw code that forms the website as the Internet user sees it. By carefully parsing the raw code, the desired information can be extracted from the mess of code that web scraping generates. In most cases, web scraping will need to compile data from multiple different pages, as a large database is often displayed across hundreds or thousands of different web pages, one for each observation in the dataset. For example, consider any of the websites that display information on homes for sale (Zillow.com, Redfin.com, Domain.com.au, Realestate.com.au, For Sale by Owner sites, etc.). Each property has its own webpage with a unique URL. To scrape the data from a site such as this, each individual page would need to be scraped, parsed, and compiled.

Repeated scrapings of multiple pages is not without its difficulties. Many websites that host large amounts of property data will issue a denial of service if too many requests for data are made from a single IP address in a short time frame. A common fix for this issue is to perform the web scraping in batches with large pauses in between. For example, if a website commonly stops providing data after 100 searches in a minute, then batch process would scrape 99 pages then wait 60 seconds and then do it again, etc. When batch processing in this manner, the data are generally appended to an existing file after each batch so as not to lose any data collected in the case of a failure.

File formats

If multiple file formats of a given dataset are available, it is best practice to collect the least proprietary and simplest file format possible. For example, if data are available in Microsoft Excel (.xls or .xlsx) or in a comma separated values (.csv) format, a researcher should collect the .csv format, even if the researcher does the analysis in Microsoft Excel.

The simpler and less proprietary format is preferred for two reasons. First, simpler formats are more likely to be easy transferred between the various software products being utilized. Nearly all database storage systems will easily accept .csv or .txt files, but many struggle to properly integrate an .xls or .xlsx format. Second, sharing data among team members, many of whom may use different tools, is much more easily accomplished with generic file formats that do not require expensive proprietary software to open and utilize. When it comes to file formats in data gathering, simple is preferred.

A caveat to this advice is in instances where data is offered in spatial and non-spatial form. Spatial data includes the relevant spatial characteristics needed to map or display the data in a Geographic Information System (GIS) or other mapping program. In most cases, the underlying tabular data is the same except the spatial data has an added layer of spatial specificity. Even if there is no plan to use the spatial data, it is wise to collect it as future iterations of the project may find it useful.

Conclusion

The process of data gathering has changed greatly over the past two decades. What was once a matter of physically collecting data is now often a process of writing code to allow one computer to talk to another. As a result, issues like licensing, use assessment, storage capacity, security, and repeated access are critical in data collection planning. Add to this the practice of web-scraping and possible legal ramifications, data collection in real estate has become as much about covering one's legal bases and process as the actual rows and columns of data itself.

6 Data management

As discussed earlier, in between gathering data and analyzing there exists a set of tasks that are crucial to the success of a research or analytical project. This chapter outlines a set of processes and procedures for efficiently managing complex and diverse real estate data prior to the Data Preparation phase – the Data Management phase of the analytical workflow. Data management is approached from the perspective of typical, large scale, and complex analysis where data is gathered from many disparate sources and needs to be combined for the purposes of the analysis. Researchers should assume that the analysis will be re-done in the future so data will be added over time, and they should endeavor to set up a process whereby new data can be added in the most efficient possible manner.

From a task-oriented point of view, there are two main activities in data management: arranging and appending data. Before we discuss these ideas, this chapter will cover some general issues regarding the best practices of data management. The chapter concludes with a discussion on the selection of data management software.

Good practice

There are a number of practical data management habits that can help the real estate researcher, whether they are a novice or veteran, with being more efficient and a better collaborator. Note that this list is aimed at the applied researcher, not a seasoned database manager, and therefore it does not include all potential data management techniques.

- *The table is the unit of management.* Each separate and unique collection of observations from each source is stored as its own table. A collection of observations is unique in the level and type of observation, as well as the source. In other words, each table should contain observations that describe the same type or class of entity – person, building, place, time, phenomena, etc. – and have the same origin. For instance, if the data are from an industrial market where there are sales transactions from source A, rental transactions from Source A, and rental transactions from Source B, all three data sets should be stored as separate tables. This holds true even if the fields are the same across the three datasets. Appending the various data sources as needed occurs later in the data preparation phase (Chapter 7).
- *Rows are observations, columns are fields.* Observations are the entities being measured; fields are the various pieces of information collected about each entity. Therefore, each row should contain one observation (and only one), as well as all of the known information about that observation. Nearly all data management software requires data be stored in a matrix format with rows as observations and fields as columns.

The main reason for this is that most data is long (more observations than fields) and not wide (more fields than observations). This is especially true in the age of big data.

- *All observations should have unique identifiers.* Every observation should have a unique identifying value, and all of these values should be in the same field. It is recommended, though not required, that this field occupy the far left hand column of the data table. Assigning a unique identifier, if one does not already exist, will assist in the joining of table, the tracking of errors, and the identification of duplicate values.
- *No blank values.* All values – every cell in the observation/field matrix – should have a value in it. If the data are missing, a value should be indicated with a designated symbol for missing data, not merely an empty cell. Leaving a cell empty could indicate a missing value or merely the non-existence of a given field, such as no garage, no agent comment, or no special circumstances in the lease arrangement. Depending on the choice of software, this may or may not be an issue. Many of the specialized data management software products will not allow for empty values and will automatically fill in the value with a chosen symbol indicating missing data. If a non-specialized software, such as Microsoft Excel, is used for data management, then there is not automatic protection against this omission.
- *Metadata and Data Dictionary.* Metadata, the data about the data, should be compiled for each data source. This should include the source of the data, the date accessed, the level or type of the observations, and other specialized information, such as the projection and coordinate reference system (if geographic data). Also, a data dictionary that describes each field should be obtained or constructed by the researcher or analyst and maintained in the data provenance.
- *All operations are recorded.* Again, as noted in Chapter 4, all the operations done in the data management steps discussed here, as well as those in the chapters that follow, should be recorded and maintained as part of the data provenance.

Arrange

The first step in the data management process is to "arrange" each data table so that it fits both within the requirements of the software as well as follows the standards of good practice, as noted earlier in the chapter. If using specialized software, some of these tasks may need to be performed prior to importing into the software product if an appropriate importing or converting process does not exist.

Orientation and unique IDs

To begin, ensure that the data are in a table format and that the orientation of the data is such that observations are in the rows and the fields in the columns. Next, check to see that the observations in each table are from a single source and represent a single type of measured entity. If not, data may need to be split into separate tables or re-arranged such that a single observation occupies each row.

Additionally, data from some sources may be unnecessarily wide in that some of the categorical fields (or variables) have been converted into a set of collective fields (Wickham, 2014). For example, take a dataset that has a list of postal codes along with the median sale price in each for the past twenty years. In many cases there will be a field for each year,

Table 6.1 Table structure example

Incorrect Table						*Correct Table*		
Post Code	*1997*	*1998*	...	*2015*	*2016*	*Post Code*	*Year*	*Median Price*
1000	$130,000	$135,000		$195,000	$200,000	1000	1997	$130,000
1001	$115,000	$117,000		$177,000	$181,000	1000	1998	$135,000
1002	$144,000	$151,500		$212,000	$235,000	...	...	...
1003	$212,000	$215,000		$295,000	$305,000	1000	2015	$195,000
1004	$125,000	$132,000		$192,000	$198,000	1000	2016	$200,000
1005	$101,000	$105,000		$160,000	$166,000	1001	1997	$115,000

making this a very wide table. The year itself is actually just a single characteristic of the observations and should be stored as a single field indicating the year of the observation. Table 6.1 provides an example where there are actually only three fields – post code, year of observation, and median price.

While the rationale for this may not seem immediately obvious, this type of table arrangement offers a very practical benefit. In cases where data are continually collected, it is much easier to simply add rows (new observations) to data than it is to add columns. Adding new columns is especially problematic if you are using analytical software packages that are script- or code-based as new columns mean changing the code, while new rows means simply re-running the script.

Locating the unique identifying field or fields may also help to determine if the table contains one and only one unique set of observations. If a unique identifying field or combination of fields does not exist, you should add one at this time. Unique identifiers can be as simple as a sequential number that relates to the observations position in the raw data as obtained from the source. Often more helpful is some value that directly relates to some of the identifying information in the observation itself. For example, if the level of observation is that of a home, then the address or the local parcel identification number (PIN) is a good choice for a unique identifier. If the level of observation were the transaction, then address is not a good unique ID because it can be duplicated if a property sold twice. In the case of transactions, a recorded deed ID or some combination of PIN and sale date might be a sensible unique ID. In the case of a property securitized as part of a residential mortgage backed security (RMBS), a sensible unique ID might be the name of the securitized trust or the pool combined with a loan number (because the same property can be securitized more than once).

If the data are part of a panel data set (the same entities observed periodically over a given time period), then the unique identifier may be defined by two (or more) fields – the observation-specific identifier and the time identifier. The example in Table 6.1 shows one such situation where a panel data set requires the postal code and the year of observation to form a unique identifier. Situations other than panel data may exist where more than one field may need to be concatenated in order to generate a unique identifier. When joining (see Chapter 7) panel data or other tables with multi-field unique identifiers, extreme caution is required to avoid spurious joins and unnecessary observation duplication.

Field names and structure

Once the table is properly oriented with observations from a single type of entity and a unique ID, the fields should be analyzed for structural integrity or correctness. First, determine whether the field names are in a format that works for the data management software as well as for additional software that may be used in the analysis part of the process. Many analytical software products both limit the number of characters in a field name and the use of special characters and spaces in the names. You should select a naming convention that is consistent, descriptive, but not too long, and works with all future software. Changes to field names will require updating the existing data dictionary (see Table 4.1 for an example) that was obtained from the data source or to begin creating one yourself. Long and thorough explanations of fields can be placed in the data dictionary; the field names should aim for succinctness.

After the fields are appropriately named, each field should be checked for consistent structure. This means verifying that the data types of all values in each field are congruent (e.g., no text or string values in a numerical column and vice versa). An example of a structural error is shown in Table 6.2. Again, many advanced software products will identify these issues for you. But caution is advised as some programs will automatically "coerce" inconsistent data types to a missing value or some other format upon import, often without notice. Here a researcher or analyst should determine if errors that are found can either be repaired with a systematic or evident fix and validate that this correction is legitimate. If the errors cannot be fixed, then the observations with these offending values should be removed and noted in the data provenance. The structural error in the "Home Size" field of Table 6.2 is one that likely cannot be fixed, whereas the error in the "Nbr of Bathrooms" can be easily remedied.

Part of the process of validating field structure is to ensure that a consistent representation of missing data is being employed. This entails confirming that there are no blank values and that all missing-ness in the data is being represented with a consistent and easily identifiable symbol or value. It is best to avoid values that could possibly be mistaken for a legitimate value, such as a 0 or –1. Depending on the analytical software to be used, these may or may not be readily identified. Most specialized data management software will employ a special symbol or data type to indicate missing values (e.g., "NA," "none," –99). This automatic handling of missing data is one of the key benefits of specialized data management software.

Table 6.2 Structural error example

Address	*Home Size*	*Nbr of Bedrooms*	*Nbr of Bathrooms*	*Presence of Garage*
7 Spring Street	1,200	2	2	0
30 Elm Street	1,700	3	2	0
108 First Avenue	*Large*	4	3	1
14 Draper Street	2,220	3	2.5	1
222 Park Blvd.	1,500	2	1	1

Append

Appending involves combining observations of the same entity from the same data source. Data management for many projects, especially those dealing with cross-sectional and/or a limited time frame, may not require any appending operations. On the other hand, for long research projects where current market trends are continuously being gathered, consistent and continual appending of incoming data will be one of the most time consuming data management exercises.

Appending is simple. As new or additional observations are obtained, they need to be added, or appended onto the existing table. Proper choice of a data management software will allow this process to be as automated as possible, with a few clicks of a button or re-running a few lines of code to complete the appending exercise. The difficult part may occur if the original data contained a number of issues with data orientation, field structure, or unique identifiers. When the data to be appended has the same problems, a systematic solution to properly arrange the data prior to appending should be developed.

Additional problems can also arise if the source of the new data has changed field types, added or subtracted fields, or made a change to unique ID or data orientation. We recommend building an automated script to check all new incoming data for arrangement compliance prior to attempting appending. This includes, but is not limited to, "assert" commands that confirm the type of values contained in a given data field.

Data diversity issues

Complex real estate analysis often requires compiling and integrating diverse data from multiple sources. While this can make for a very rich collection of data and meaningful analytical results, it does bring with it a number of additional issues during data management. Two of the most common of these are temporal inconsistencies and spatial misalignment.

Temporal inconsistencies

Real estate markets, as well as the physical structures themselves, are dynamic. As a result, even when diverse datasets are gathered at the same time, the data themselves may represent observations that represent vastly different time periods. A common example is the use of government collected census data to bolster housing price models. While the transactions used in the analysis may be quite recent, the census data itself may be up to a decade old and may no longer represent the "facts on the ground." Likewise, many pricing model studies that use a long time frame of transactional data may end up joining those transactions to the physical characteristics of the properties as they exist in the most recent data that are available. When this occurs, it is possible that the data represented may not align with the true condition or land use at the time of the sales transactions. For example, data gathered from multiple listing services (MLSs) or other industry sources may have property characteristics data that are joined to the transaction information at the time of the listing or sale. This would avoid the issue of temporal inconsistencies. MLS data, however, is entered by agents and can have varying information, though it is temporally correct.

There is no easy fix for issues involving temporal inconsistencies. If available, historical data for one or more tables may be gathered to better align with some of the dated data. In the case that historical data are not available, the issues of temporal inconsistencies should be duly noted in the limitations section of the study. As most real estate data are observational, as opposed to experimental, researchers in our field must often "play the hand they are dealt" and make the best of the available data. In any case, the meta-data for each data table should indicate both the time of data collection and the temporal span of the observations themselves.

Spatial misalignment

Like issues of temporal inconsistencies, the areal boundaries used by various data sources may not exactly match. For example, different industry sources may utilize different "submarket" boundaries when classifying sales or rental observations. These boundaries may also differ from "neighborhood" boundaries used by local government, from official national census areas, or from statistical agency collection areas.

Spatial misalignment issues are most prevalent and most difficult to manage when dealing with spatially aggregated data. A simple approach to minimizing impacts from spatial misalignment is to always gather data at the most disaggregated level possible as it is easy to aggregate but very difficult or impossible to disaggregate. Geographic Information Systems (GIS) can also play a role in helping to resolve issues of spatial misalignment. Like temporal inconsistencies, there is no easy and certain solution; researchers and analysts need to be aware of the potential for these problems and communicate how they were overcome.

Choosing a software

As discussed in Chapter 3, there are many software options for working with real estate data. Each project, study, report, or product is different and, as a result, there is no one solution to all of the needs that researchers or analysts may have. The larger and more complex the data, the more likely a specialized software, or a combination of specialized software products, will be needed to best perform the task(s) at hand. This often means expanding beyond spreadsheet-based products. In choosing the correct software or suite of software products for data management, it is important to ask a number of initial questions:

1 *How many tables (data sets) will I have, and will I be combining them?*
Projects that have many diffuse data sources and require the joining of these data benefit greatly from data management software that stores the data and manages linkages through relational databases. Microsoft Access is the simplest and most common relational database management system. Other larger and more functional systems (some for a fee and some with no fee) also exist.
2 *How big is my biggest dataset (in terms of observations and fields)?*
Very large data, either tall (many observations) or wide (many fields), may also require more advanced data management software. It is advisable to get an estimate of the size of the largest data table that may be required before choosing a product.
3 *Will I be updating my data frequently, occasionally, or not at all?*
Data that need to be updated frequently, as is often the case with real estate market data, require a product that can easily add new observations (appending) to the existing tables. Failure to account for this fact when choosing software can greatly increase maintenance costs.

4 *Am I limited by cost or is user support more important?*
There are many viable options for large data projects that are free and/or open source. However, many of these products offer little in the way of specialized technical support. Understanding the resource limitations and support needs of your team should be considered in choosing a data management software.
5 *Who other than my team and I may be using the data in the future?*
If the users of the data are a small known group, then the choice of software can be limited to what the current team is comfortable with. However, if future users may reside outside of the initial group, then more standardized products or those with better integrative capabilities may be a wiser choice.

In short, the choice of data management software is a critical decision. It is not necessarily an irreversible one, but changing in the middle of a project will cost precious time and potentially financial resources, so it is best to evaluate options and make the best possible decisions from the outset.

Conclusion

In this chapter we discussed the data management phase of the analytical workflow. Data management covers those operations and tasks that take the raw data from the source(s) and bring them together into a single location. Along the way the data are arranged into a proper tabular format that lends itself to the subsequent data preparation phase. As such, data management is concerned with the basic, structural features of the data (the rows, the observations, field names, data types) and with the metadata (the temporal and spatial-ness of the data). Investigations and operations on the data values themselves are done in the next phase: Data Preparation.

Setting up an efficient and user-friendly data management system or process is critical to any medium to large-sized real estate analysis project. A key component is designing a process that allows for new or additional data to be easy added as new observations are created in the market. Following good data management practices will assist researchers in both understanding the strengths and weaknesses of their data as well as in preparing the data for the ensuing analyses.

Reference

Wickham, H. (2014). Tidy data. *Journal of Statistical Software*, *59*(10), 1–23.

7 Data preparation

Like data management, the preparation of data prior to analysis is a critical, yet often overlooked step in the research process. In simplest terms, data preparation is the link between the managed data and the initial analytical processes such as exploration and cleaning. Part of the difficulty in defining data preparation is that the exact set of steps required differs for nearly every research project, depending on the research question, the intended form of the output, and the quality of the raw data.

As a result of the inherent idiosyncrasy in the data preparation step, a standard vocabulary is lacking. Not having a standard set of terms means that communicating one's own or trying to understand the data preparation undertaken by another party can be difficult under the best of circumstances. Below is one set of terms to describe the various operations generally contained in the data preparation phase. These may or may not exactly match the language in other fields or those used by other researchers. For this reason the types of actions and tasks that each involves are outlined so as to best allow for translation into another lexicon of data preparation.

There are three large categories of tasks that make up data preparation. These tasks are presented in the general order in which they may happen; however, most data preparation exercises are an iterative process, and a researcher may return to earlier steps many times as they work through the data. Further, to place data preparation in the analytical workflow, it falls after data gathering and management but before exploratory data analysis and data cleaning.

It should be noted that within industry and many research teams, "data cleaning" is used as a broad term to describe the data preparation process. This term is avoided at this stage as it does not adequately describe the breadth of tasks needed to prepare data. Data cleaning should instead be considered a finalizing process that, together with exploratory data analysis, removes observations that are suspected outliers, contaminants, or insidious errors. More on the data cleaning process can be found in Chapters 8 and 9. In short, we see data cleaning as a specific process driven by statistical tests and visualization – and aided by domain-specific knowledge – that targets the removal of a small number of observations from the data. On the other hand, data preparation, which is guided by the state of the data, the research question at hand, and the intended methodology and outcome, involves a much wider set of tasks.

Within data preparation, there are three core processes (Figure 7.1):

1 Transform
2 Integrate
3 Filter

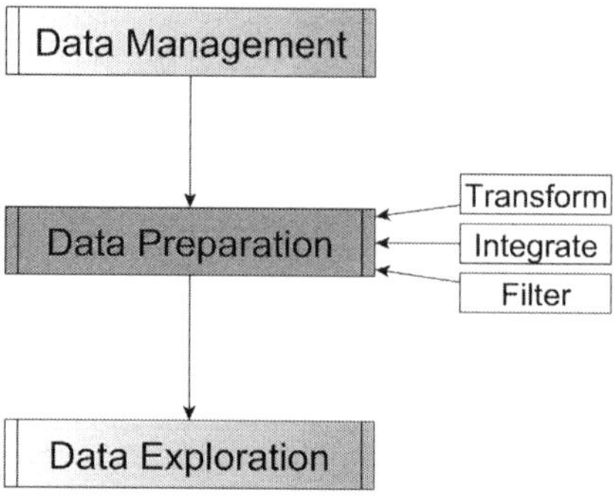

Figure 7.1 Data preparation in the workflow

Again, the data preparation steps for an individual project may not include all processes, and complex data preparation may not be a linear process. It is highly likely that some combination of these steps will need to be done iteratively and potentially out of order.

It is important to note that the operations discussed below are not applied directly to the raw data or even the "managed" data from Chapter 6. When data is changed, temporary tables are created to hold the information so that the researcher can always backtrack to the original in the case of a mistake or the need to redo an earlier step. At the end of the data preparation process the researcher will want to save the prepared data table(s). In other words, researchers should always maintain a copy of the data from previous steps, and work on temporary or in-memory data as they proceed through the data munging steps.

Transform

Transforming the data is usually the first stage in the data preparation process. Transforming data involves operations on the fields (columns) of data. At this juncture, no observations (rows) are removed or individually impacted. A common purpose of transforming the data is to prepare individual data sets for integration with other data, a process covered in the next section.

A number of actions can be performed on the data fields. The simplest and most obvious is the removal of a field. Fields that are not needed for the analysis can be safely removed from the working data table(s) at this time. A researcher may want to remove unnecessary fields for two reasons. First, if the data is very large, removing excess fields can free up memory, especially in programs that do most of the analysis in working memory (RAM). Second, and more a matter of preference, it can be helpful to remove unnecessary fields to aid in easy visualization of the tabular data during the processes that follow.

A number of operations more complex than removal may also be done at this stage. These are combining, converting, and rescaling. As each of these operations take one field and change it into another, they can collectively be referred to as "mutating."[1] Combining involves the merging of two or more fields in to a single field. Combining is common in instances where the raw data includes a segregation or granularity of data that is unnecessary or a hindrance to answering the research question(s) at hand. For example, if the data has home size information broken out room-by-room, it may be wise to combine this figure into a single field noting the aggregate home size. The redundant, original fields can then be removed. Note that combining is usually performed on fields that have continuous data types.

Combining need not be just the simple addition of two or more fields. Many regression model specifications utilize interaction variables, where the impacts of the product (multiplication) of two fields are entered into the model in order to observe the conditional or combined impact of the two on the dependent variable. Creating interaction variables in the dataset at this stage can also be considered combining, and it falls within the transform stage.

Converting is a process that is closely related to combining, but it deals with categorical or ordinal data. The process of converting can either create more fields from an existing field, *expanding* (Table 7.1); aggregate a number of existing fields into a single field, *collapsing* (Table 7.2); or change the number of categories in a given field, *re-categorizating* (Table 7.3). For example, if a dataset has a field containing a measure of the building

Table 7.1 Expanding a field

Existing Table		*New Table*					
Address	*Home Condition*	*Address*	*Poor*	*Below Average*	*Average*	*Good*	*Excellent*
7 Spring Street	Average	7 Spring Street	0	0	1	0	0
30 Elm Street	Good	30 Elm Street	0	0	0	1	0
108 First Avenue	Below Average	108 First Avenue	0	1	0	0	0
14 Draper Street	Poor	14 Draper Street	1	0	0	0	0
222 Park Blvd.	Average	222 Park Blvd.	0	0	1	0	0

Table 7.2 Collapsing fields

Existing Table						*New Table*	
Address	*Poor*	*Below Average*	*Average*	*Good*	*Excellent*	*Address*	*Home Condition*
7 Spring Street	0	0	1	0	0	7 Spring Street	Average
30 Elm Street	0	0	0	1	0	30 Elm Street	Good
108 First Avenue	0	1	0	0	0	108 First Avenue	Below Average
14 Draper Street	1	0	0	0	0	14 Draper Street	Poor
222 Park Blvd.	0	0	1	0	0	222 Park Blvd.	Average

Table 7.3 Re-categorization of a field

Existing Table		*New Table*	
Address	*Home Condition*	*Address*	*Home Condition*
7 Spring Street	7	7 Spring Street	3
30 Elm Street	11	30 Elm Street	4
108 First Avenue	5	108 First Avenue	2
14 Draper Street	2	14 Draper Street	1
222 Park Blvd.	8	222 Park Blvd.	3

quality on a scale of 1 to 5, expanding this would convert this single field to five individual fields of binary data types, each indicating whether or not the observation has a quality rating of poor, below average, average, and so on. Collapsing, conversely, would take a set of five binary data fields and collapse them into a single field with the five different categories or levels. Finally, factors or levels within a field can be re-categorized as well. Often a field will have too many different levels for analysis, say if there were 15 levels of quality instead of 5. Re-categorization could condense these 15 into 5 levels. Note that re-categorization almost always works in going from many to a few, unless other fields are used to help disaggregate existing categories into a finer specification.

Rescaling involves systematically changing the scale or precision of the data in the field.[2] A simple example would be data that is in square feet but needs to be converted to square meters or vice versa. The precision of the data can also be altered. If the data were measured to the .0001 square meter, that level of accuracy may not be needed, and the field could be rounded to the nearest square meter. Rescaling operations are generally limited to continuous numerical data types; changes to the categories or length of categorical or textual data usually fall under converting.

The final transformation process is that of labeling. Labeling is critical to understanding, identifying, and classifying observations in large datasets, especially property level real estate data. Labeling is similar to some of the converting processes, but it differs in that the creation of new data fields is driven primarily by some external phenomenon that has domain significance. Often numerous fields may need to be referenced in order to properly assign a label, or occasionally some measure of subjective judgment or heuristic is used to determine how the existing data will be used to create the new label. Converting, on the other hand, is a simple and formulaic conversion of one field to another.

As an example of labeling, foreclosed, distressed, or short sale transactions may not be considered valid observations if the study is not aimed at addressing phenomena surrounding these types of transactions. Non-arms length sales may be identified and labeled by using the sale price field in concert with sale date, a sale warning, and a buyer name field, for example. Again, while labeling often follows a logical process that depends on one or more existing fields to create a new field, converting merely expands, collapses, or re-categorizes existing information.

It is important to fully understand the data source before labeling the data to determine if there are any restrictions on the dataset. If the data will be assembled (combined together by row) in a future step, it can be useful to create a field that indicates the source of the data so that future issues can be traced back to their original source.

Integrate

The second step is data integration. Integration refers to the combination of diverse datasets. Integration can occur in one of two ways, assembling or joining. Assembling involves combining observations (rows) from different data sets that have the same level or entity of observation. Visually, the data are added on top of each other, thus assembling is a "vertical" integration of the data. Joining, on the other hand, involves adding fields from two or more tables based on a common field between the two. In a join, the number of observations does not increase, but the number of columns will increase. Hence, it is a "horizontal" integration of data.

Assembling often involves integrating data from different sources that describe the same phenomenon. For example, if the data have retail rental transactions from two different industry sources, these would be assembled to provide a complete list of all the retail lease transactions. Mechanically, assembling is similar to "appending" discussed in Chapter 6. However, assembling the data involves different sources, whereas appending data involves the same source. Prior to assembling, data often require transformations to the field names to ensure that fields match.

One of the biggest challenges to assembling data is the lack of field standardization that often occurs when different data collectors employ diverse data collection or management techniques. As an example of a data collection standardization issue, some data sources will count foyers, open spaces, covered porches, and basements in the square footage of a home while others may not count them. When assembling these two data sources, the lack of standardization will likely cause problems. More serious standardization issues can occur in relation to data fields that are somewhat subjective, such as building quality or condition. It can be very difficult to ensure that diverse data collectors are using the same rating system.

Another type of standardization issue occurs when different data sources use different scales or data types to represent certain fields. Table 7.4 gives an example where two different data collectors are using different scales, precision, and data type in their data. In the quality field, there is a difference of data type. In the case shown, it may be possible to convert between the two if both are recorded on an ordinal scale with the same number of levels. For the bathrooms field, the differences between the two sources are an issue of precision. The data from Source 2 is more precise in that it includes fractional bathrooms, while Source 1 only lists full (or rounded) bathroom counts. In situations like this, the entire field should be rounded to the less precise of the two, the Source 1 option. Finally, the garage field shows a difference in scale. Source 1 measures garages by their size, while Source 2 measures by parking capacity. Here some conversion factor may have to be applied to transform one of the sources to the other. In choosing which to convert, the one with more variation (Source 1) should be converted to the once with less variation (Source 2).

Horizontal integration is often referred to as "joining" in computer science or database management lexicon. To join means to link two tables together based on a common field. In at least one of the datasets the common field should be a unique identifier. Joins can be of one of three standard formats: inner, outer, and uneven (left and right).

1 Inner: Only observations that match in both tables are maintained.
2 Outer: All observations from both tables are maintained, regardless of match.
3 Uneven (Left or Right): All observations from one table are maintained, while only those that match are kept from the other table.

Table 7.4 Between source standardization examples

Address	*Source*	*Quality*	*Bathrooms*	*Garage*
7 Spring Street	1	Average	2	300 sf
30 Elm Street	1	Good	3	480 sf
108 First Avenue	1	Below Average	3	440 sf
14 Draper Street	2	3	2.25	1 space
222 Park Blvd.	2	4	3.25	2 spaces
1 Ace Court	2	2	2.75	2 spaces

The integration process may involve the joining of multiple data tables, potentially joining by different fields and in a different format. Additionally, note that inner joins may hide Errors of Join Matching and should be done carefully.

Joining essentially "matches" a row from one dataset to another and then adds all of the data from the two datasets together into a row. In many cases one of the tables will have significantly more observations than another table. Imagine a dataset with 100,000 sales transactions, each with a postal code identifier. There may be 300 different postal codes in the data. If information from a postal code table that includes median income and average household size is joined to the transactions data, the action is, in essence, joining a table of 300 observations to one of 100,000 observations.

Even when both datasets have a common field, such as address, some difficulty may remain in finding direct matches if the fields are formatted differently. Most data management software will only make join matches on exact matches, and a single comma or space out of place can cause a failed match. Fortunately, many analytical programs such as Microsoft Excel, R, SAS, and Stata do allow for fuzzy matching, a technique that can make matches on similar, but not exact, values.

Another option may be available when the data have spatial attributes. If one of the datasets represents point observations (a specific point in space) and the other dataset is areal (a defined area of space or a polygon) then Geographic Information System (GIS) software can be used to "spatially join" fields from the areal dataset onto the point dataset. Note that areal-to-areal joining may be possible but can require complex GIS operations.

Filter

The third and final step in data preparation is filtering. Filtering removes observations (rows) from the data. At this stage, observations are filtered out for one of three reasons. The filtering can occur in any order, but may be more efficient in a certain order depending on the software used in the data preparation process.

Relevancy

The first consideration in filtering the data is relevancy. Throughout the data gathering process researchers will often collect as much data as possible that might help answer the question at hand. Next, in the data management phase, this data is brought into a proper, workable format. As the applicability of all observations may not be known until after integration, the researchers proceed with transforming and integrating all of the original observations. Thus, at this stage in the data preparation it is possible that researchers have collected, stored, managed, and prepared observations of such a type that are not useful in the forthcoming analysis. Here is where that is remedied.

For example, imagine that researchers are conducting a study that focuses solely on the retail rental market in a particular metropolitan area like Chicago, but the data source A also contains observations from Milwaukee and Detroit. Source B has only Chicago data, but it also includes observations from the office market. At this stage researchers would remove the Milwaukee and Detroit observations, as well as the office observations from the data. Often the labeled fields can be very helpful in determining the relevancy of a given observation. As with all operations in the process, the rationale for these removals should be noted.

Duplicates

The second filtering task is to remove duplicate observations from the data. Duplicates at this stage in the data preparation process can arise in a number of ways. The most common is through the integration process. Joining or assembling data from two or more data sources often results in duplicate observations. Duplicates caused by joins are often "exact" or "mechanical" duplicates, meaning an observation (or many of them) is simply duplicated due to the matching process of the join. These can usually be remedied by examining the unique identification numbers to see if any repeat observations are in the data.

More difficult to recognize and correct are duplicates created through assembling two or more datasets. In these cases the duplication is often not readily recognizable by the unique identifiers, and it needs to be evaluated based on other fields. As an example, if two sets of retail rental observations are assembled from two different industry sources, it is possible, in fact likely, that some of the individual observations will be located in both of the sources. If the unique identifiers do not reference the key differentiator fields, then it is not entirely obvious which observations are duplicated. To find and remove these duplicates, it may be necessary to run a number of tests or IF statements (a logic equation; if, then, else) to see if any observations reference the same address with the same date and for the same price. If so, they are likely a repeated observation and can be removed.

A more difficult situation can arise when duplicated observations from assembled sources have conflicting field values. If it can be confirmed that the observations are duplicates but some of the additional (non-identifying) fields do not match, then the problematic data must be reconciled. For example, there are a number of duplication inconsistencies shown in Table 7.5.

As first step, the source should be categorized to see if either is the "super-set" of data. What this means is that if one source, Source 1 in this case, contains all observations, and Source 2 only contains some subset, then Source 1 should also be the default data set. Indeed, in the case where one dataset is a superset of the other it may not be advisable to assemble the data at all. If neither source represents a superset then a reconciliation process is warranted. If the data values that do not match are numeric, an average, or some other measure of central tendency, can be taken.

If many such inconsistent duplicates occur, then it should be determined if the inconsistencies are systematic or random. If systematic inconsistencies occur – say all of Source 2's values are 10% higher than Source 1 – then a correction can be employed to all duplicate and non-duplicate values to ensure that the final assembled data is consistent. If the inconsistencies are random, then the researcher must make a determination which data set or values are more likely to be correct. Large discrepancies may lead to the removal of the observation all

Table 7.5 Duplicated inconsistencies

Address	*Source*	*Quality*	*Bathrooms*	*Garage*
7 Spring Street	1	Average	2	300 sf
30 Elm Street	1	Good	3	480 sf
108 First Avenue	1	Below Average	3	440 sf
7 Spring Street	2	Good	2	330 sf
108 First Avenue	2	Below Average	4	440 sf

together, whereas small discrepancies may be able to be individually or systematically reconciled without a large impact on the subsequent analysis and reliability of the results.

Failed matches and missing data

Missing data is a serious issue that plagues many datasets. The missing-ness is usually from one of two processes. First, the raw data may contain missing values or fields due to poor data collection processes or from the source's own data preparation process. Second, the integration operations performed above can create missing data from a failure to match on the joining fields.

Issues of the former – those originating in the raw data can require complex statistical methods to properly assess and correct and, as a result, are discussed more fully in Chapter 9. Missing data deriving from failed matching in the join process can be handled at this stage in the data munging process.

The first step is to confirm that the join has occurred properly. Many software products will not accurately join two data tables if the join fields are stored as different data types. For instance, if postal codes are stored as a numeric value in one table and a categorical in another, a join may not occur or may occur erroneously. The next step is to determine the extent of the failed joins. A failure to join a large portion of the data would suggest that the existing data may be insufficient to answer the research question at hand or significant additional data collection needs to occur.

If there are only a small number of failed matches, every effort should be made to determine if the failed matches are a result of incomplete data collection in either table. For example, if a few postal code observations are missing, this will result in a small number of failed matches. If these few missing observations can be added to the data, the failed matches can be avoided. If/when it is determined that all of the data have been properly collected, then the remaining failed matches are usually the result of temporal or spatial inconsistencies in the extent of the data (the observations of one data table exist outside of the time or space covered by the data to which it is being joined) or are simply additional irrelevant observations that have gone undetected in the earlier labeling and filtering steps. In either case, these observations can be filtered out, as the missing-ness that results here is systematic and related to the research design.

Conclusion

In this chapter we covered the three main operations of the data preparation phase: transforming, integrating, and filtering the analytical workflow. These steps occur after the data have been managed but before exploring and data cleaning. They are often completed iteratively and can happen many times over the course of the analytical workflow for a large and complex real estate research project. Smaller projects may result in a more streamlined workflow, but even that data can be subject to repeated changes through the course of the research.

Notes

1 Like much of the language used to describe the data preparation activities, we refer to Hadley Wickham's very successful dplyr package and associated lexicon: https://github.com/tidyverse/dplyr.

2 Note that rescaling of continuous or discrete numeric data is similar to the re-categorization of ordinal data.

Part III

Modeling and analysis

8 Exploratory data analysis

In most situations a real estate researcher or analyst will be working with secondary data: i.e., data collected by other parties. Users of real estate data may not be as familiar with the structure, overall characteristics, and potential shortcomings of the data that they are working with to the same level as might be, for example, an archeologist who has personally collected and hand-entered the data for her study. Considerable effort, therefore, should be made to "get to know" the data prior to making key decisions on choosing analytical techniques such as model specification. The familiarization process is known as exploratory data analysis (EDA) and includes traditional techniques such as scatterplots, as well as all manner of mapping and some forms of statistical and spatial analysis.

This process is distinct from data management and preparation. In the prior steps the general concerns were about the structure of the data, its completeness, its structural relationships (joins), and its relevance to the project. From this point forward the focus will turn toward the actual values of the data itself.

Exploratory data analysis stands out from much of what has traditionally been considered standard statistical practice. In the standard approach of hypothesis testing, model fitting, and diagnostic testing, the analyst is setting out to confirm (or deny) a given hypothesis or set of hypotheses. This type of analytical work is generally referred to as "confirmatory" (Behrens, 1997). Exploratory data analysis, on the other hand, approaches the question of data analysis with a blank slate, allowing the research to, in the words of Gelman (2004), "Find patterns in the data beyond the expected." In other words, EDA encourages analysts to open their mind to possible trends, relationships, and features of the data that may not have been evident due to existing theory, domain knowledge, or past work on the subject. In essence, EDA requires the analyst to play the role of a private investigator or detective, as opposed to a focused scientist. Exploratory data analysis is as much an attitude as a method (Behrens, 1997).

Since the work of Tukey (1972, 1977), the importance of exploratory data analysis to the overall analytical workflow has slowly grown. Much of this increased realization is due in part to the improvements in computation and graphics that now allow for nearly all statistical software to produce the metrics and graphs required for EDA (Gelman, 2004).

This chapter will discuss the techniques for exploring the data. Many of these same techniques are also used to identify discordant values (outliers and errors), and in practice these two operations may be done in concert. For the sake of clarity, the discussion of identifying, labeling, and handling discordant values is saved for Chapter 9.

Why EDA?

The general rationale for exploratory data analysis is that it helps analysts get to know the data. If the analyst personally collected the data and intimately "knows" the data, is EDA necessary? What does "get to know" mean exactly? The answer to the former is "yes": even if they collected the data, they should still do EDA. To the latter, following rationale is provided to support the importance of EDA.

The simplest reason for performing exploratory data analysis is that it provides a verification of previous data preparation activities. If, during the data management or preparation steps, data errors, instances of failed join matches, or irrelevant observations were not identified, these errant observations will likely show up during the course of EDA. In this manner, EDA functions as a validation of the initial data preparation process.

Next, EDA will help to identify patterns and distributions in the data. These insights are important for three reasons. First, possessing a better understanding of the basic "shape" or "form" of the data can help to redefine, or sometimes completely re-design, the research question. For example, a researcher is engaging in a study that initially set out to examine the impact of waterfront access on office rents. Upon examining the data, the researcher finds that only one of the 87 office buildings in the sample actually has waterfront access. The sample size here is too small to determine any statistical difference between offices with and without waterfront access. However, in the course of EDA the researcher finds that 23% of the leased spaces had some form of water view. With a slight change in the research question (analyzing the impact of view as opposed to access), the overall project can be continued and the likely results will be more meaningful to real estate practitioners and not simply an academic exercise.

Second, gaining an understanding of patterns and distributions in the data form the critical groundwork for outlier identification and/or missing data imputation (discussed more fully in Chapter 9). Once the basic shape of the data and the general relationships between fields are known, the selection of methods, tests, and procedures to handle outliers and missing data can be made more accurately and efficiently. It is a good idea that during the EDA process, the analyst or researcher does not remove or clean any data observations, but rather that the EDA functions as an input to or background information for the final data cleaning exercise.

Third, by identifying patterns and distributions in the data, EDA can assist the analyst in choosing modeling specifications and parameters. If, through EDA, it is determined that the distribution of the dependent variable, say retail foot fall in regional malls, follows some distribution other than a normal distribution, then the choice of more complex statistical models or a transformation of the dependent variable can take place prior to the actual modeling process. Similarly, if the EDA process shows high levels of spatial autocorrelation in the variables, the analyst will know ahead of time that certain assumptions in basic statistical models are likely to be violated and that a more spatially-explicit method of analysis should be chosen. EDA is an extremely important step in model selection and validation because complex models do not yet have statistical measure that completely and sufficiently explain relationships in the data (Gelman, 2004).

Finally, the EDA process can generate new insights on the current research question or even provide new ideas for future research. By examining all dimensions of the data, often in combination with other fields, relationships that were previously undiscovered or ignored can be uncovered. While not all relationships are meaningful, one should be on the lookout for potential new ideas during the EDA process. For example, in the water access/water view example, we may uncover some interesting patterns during the mapping of our dependent variable which seem to indicate much higher rents along busy streets. While this may at first seem counterintuitive as traffic can be a negative amenity, we may find that a bus rapid transit (BRT) line runs on these streets and offers much better access to employees of those offices. This may lead us to do one or both of two things: (1) incorporate distance to transit stop as a variable in our model to control for this influence; and/or (2) conduct a future study on the impact of BRT on office rents. The potential importance of BRT may not have been evident without the process of mapping and EDA.

The EDA process

EDA is, in short, a familiarization process by which the data are examined to determine their general characteristics. These characteristics may include, but are not limited to, central tendencies, distributions, moments, ranges, spatial and temporal attributes, types and number of clusters, and relationships to each other, both between observations and between fields.

As the human eye is much more adept at finding visual patterns than at scanning and analyzing a column of numbers, EDA often involves the use of data visualization techniques. EDA processes or methods, therefore, can be divided into visual and statistical (non-visual). Many of these visualization techniques also produce specific numeric metrics and can be considered both visual and statistical. As a result, some of the exploratory process can rely on subjective human judgment. However, many processes or tests have associated objective metrics or measures associated with them.

EDA techniques can also be categorized by the number of fields that are examined. Univariate techniques focus on a single field and are concerned with the size, dispersion, or other feature of the values in that field, irrespective of the other data fields. Multivariate methods, on the other hand, consider the importance of other fields, and are also frequently concerned with the relationship between values in different fields. Often, but not always, univariate techniques are employed first, followed by the more complex multivariate processes.

Central tendencies and distributions

Central tendencies are the most common and, often, the first and easiest metric examined in EDA. A central tendency is a single number that best describes the mass or bulk of the data. As such, central tendencies are univariate measures. An average (mean) of a set of numbers is a central tendency, as is a median or mode. While the average (mean) sale price in a neighborhood or the average retail lease rate in a shopping center does not provide all the information on those particular fields, it does, in a fast and easy manner, give the analyst an immediate sense of what is common across that field.

The type of the data field in question usually determines which of the three common measures of central tendency to utilize. Numeric data will use a mean or median, while mode is the only available metric for categorical or textual data. Also, different measures of central tendency have different biases given the overall distribution of the data. As the distribution of the data become more highly skewed, the mean and median of the given field will grow farther apart. Table 8.1 provides examples of the three types of central tendencies across a variety of data types.

Table 8.1 Examples of central tendencies

Data Type	*Sale Price*	*Full Baths*	*Building Quality*	*Police Beat*
	Numerical – Continuous	*Numerical – Discrete*	*Numerical – Ordinal*	*Categorical*
Mean	$589,854	1.50	7.61	*NA*
Median	$508,000	1	7	*NA*
Mode	*NA*	1	7	U3

Table 8.2 Example Dispersion Metrics

Data Type	*Sale Price*	*Full Baths*	*Building Quality*	*Police Beat*
	Numerical – Continuous	*Numerical – Discrete*	*Numerical – Ordinal*	*Categorical*
St. Dev	$380,020	0.68	1.05	*NA*
MAD	$220,166	0	1.48	*NA*
IQR	*$378,000 to $690,000*	1 to 2	7 to 8	*NA*

Along with a central tendency, an early stage of EDA involves examining the distribution of the data. Distribution measures or metrics explain the variability in the data and can offer a more complete picture of the extent of the data itself. There are multiple measures of distribution or dispersion, all of which are univariate. The most common is the standard deviation. In simplest terms, the standard deviation offers a measure of how much the values in a given field differ from the mean or how dispersed they are from the mean. A high (low) standard deviation means high (low) dispersion in the data. Other commonly used metrics are median absolute deviation (MAD) and the IQR (the difference between the 25th percentile and 75th percentile values). The standard deviation, MAD, and interquartile range are all used for numerical values and are most meaningful with continuous data. There are few, if any, standard tests for dispersion in categorical data (Table 8.2).

Several of the aforementioned measures represent what are known as "moments" of the data. Moments summarize some feature of the distribution of that data and are most applicable on numerical continuous data. The mean or average is what is known as the first moment, and the variance (the square of the standard deviation) is the second moment. These first two moments are the most commonly used metrics to describe continuous data.

The third moment is referred to as skewness and indicates whether the data shifts to the left or the right of a standard, normal (bell-shaped curve) distribution. Most real estate data, such as prices, rents, sizes, etc., exhibit right-skewness. Finally, kurtosis, the fourth moment, measures to what extent the data are concentrated around the mean or evenly dispersed when compared with a hypothetical normal distribution. Depending on the research question, these additional moments (3rd and 4th) may be helpful in assessing the data.

Collectively, the simple measures of central tendency and the first few moments of distribution are often referred to as descriptive or summary statistics. Occasionally, the list of median, first quartile, third quartile (first and third quartile are often collectively referred to as the interquartile range (IQR)) and minimum and maximum are referred to as the "five-number summary" (Peng, 2016). Most analytical packages or languages offer functions to easily calculate these basic figures.

Visual approaches

The most common visual approaches for univariate EDA of continuous variables are box plots and density plots. Box plots indicate the mean, interquartile range, and occasionally the minimum and maximum values, via a vertically oriented plot. Boxplots are useful to visualize the range of the bulk of the data, (those within the box, or the IQR) as well as potential extremes. Figure 8.1 offers an example of a boxplot, showing the distribution of home sale prices within the City of Seattle over the 2011 to 2016 time period. In Figure 8.1, the thick line starts at the 95th percentile, suggesting that 95% of homes sold for less than $1,150,000 over this time period.

A density plot is a substitute for a boxplot. A density plot may be more appropriate if the data has an unusual shape, such as a bimodal distribution. Density plots show the frequency of values at each point in the distribution, and therefore are better able to aid the researcher in fully visualizing the mass of data. Unlike boxplots, density plotting functions in most software packages lack markers of the location of the summary statistics (mean, median, IQR, etc.) on the graph itself. As a result, density plots are often used in combination with summary statistics. An example of a density plot using the same sales price data from the boxplot in Figure 8.1 is shown in Figure 8.2.

For categorical, ordinal, or discrete data, boxplots and density plots are not used to visualize distributions. Instead, for these data types, histograms are used for the same purpose. Histograms are more similar to a density plot, except they offer discrete categories along the horizontal axis, not a continuous value. An example histogram showing the count of bedrooms in the sales data can be seen in Figure 8.3. As can be seen in the histogram, 3-bedroom homes are by far the most common in this dataset, followed by 4- and 2-bedrooms dwellings. Very few properties have fewer than two or greater than five bedrooms.

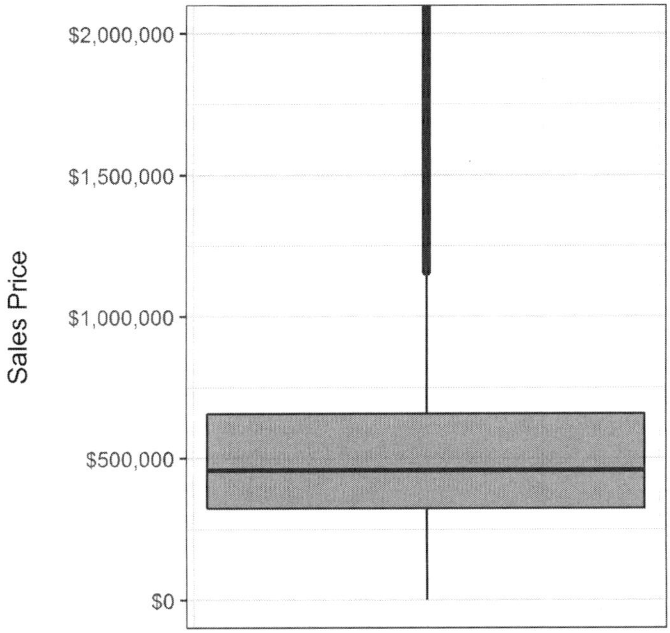

Figure 8.1 Boxplot of sale prices

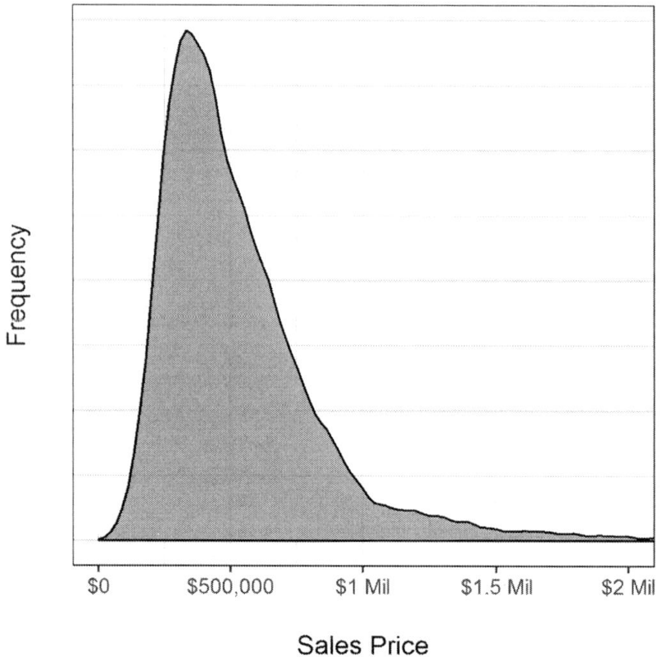

Figure 8.2 Density plot of sale price

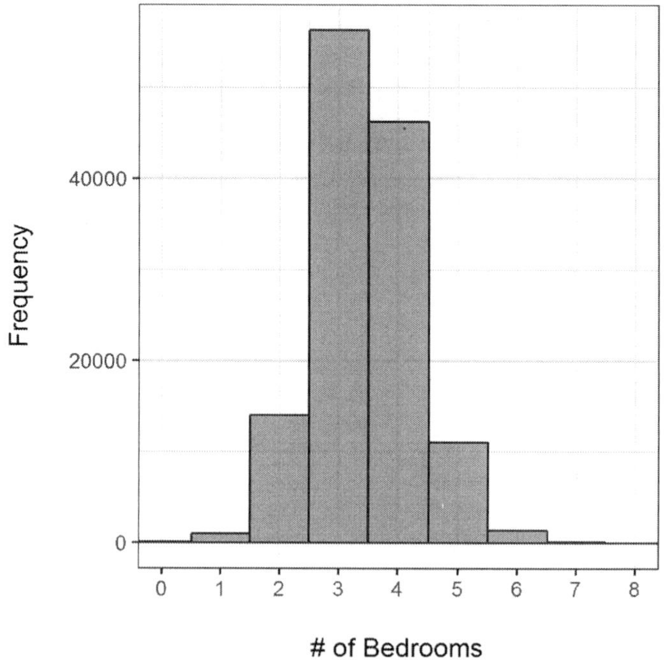

Figure 8.3 Histogram of bedrooms counts

Stratification

Stratifying a given dataset by a particular field and then applying univariate EDA techniques to the observations in each of the strata is a common method of exploring the data in a deeper way. The stratification and re-analysis process can help determine if various subsets of the data are alike enough to be considered in the same analysis or if they should be broken out in the tests and models that follow. Stratification is most easily done when the data are stratified by a discrete or categorical field with a limited number of categories, values, or levels. Data can be stratified by continuous data; however, bins or groups will need to be applied first.

As an example, Figure 8.4a shows side-by-side box plots of the distribution of sale prices split by structure type. Detached homes are shown on the left and attached townhomes

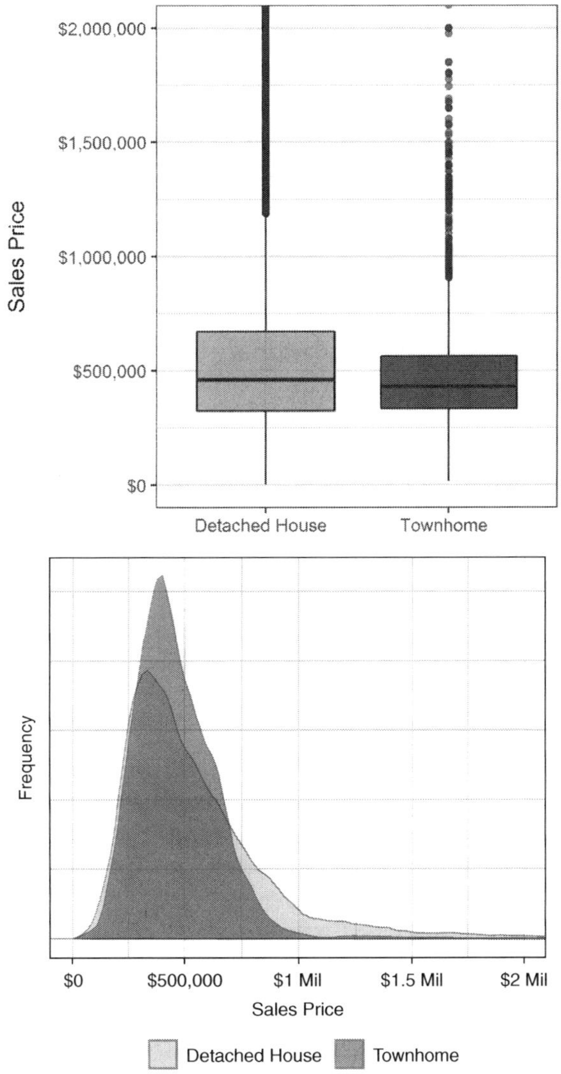

Figure 8.4 Boxplot and density plots of sale prices: detached homes versus townhomes

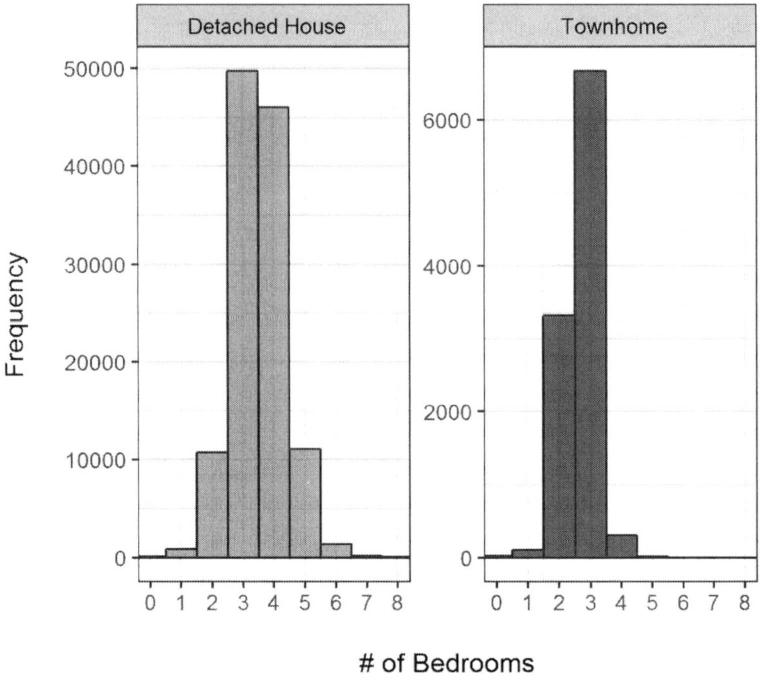

Figure 8.5 Histogram of bedroom counts: detached homes versus townhomes

on the right. The mean for detached houses is greater than that of townhomes, as is the overall distribution of prices. In the stratified density plots in Figure 8.4b, we see that the bulk of both types of sale prices range from about $300,000 to $700,000. However, there is a much larger set of the sample of detached houses that sell for $700,000 or greater.

Stratifying the data by structure type and looking at bedrooms counts with a histogram (Figure 8.5) shows that, in general, detached homes are larger than townhomes, as would be expected. Most townhomes have three bedrooms at the most, whereas there are a large number of homes with four bedrooms and some with five or more bedrooms. What this stratification shows is that if an analyst finds an interesting relationship regarding large bedroom homes in future analyses, he will know that these findings will only apply to detached houses and not townhomes.

Stratification and side-by-side display take advantage of the human eye's ability to discern differences from like presentations of sets of data. Tufte (1990) refers to this as "using small multiples," whereby a similar structure is set up and different data or subsets of data are examined in these same structures to highlight differences. In many software programs, small multiples can be created easily and efficiently with a few commands or clicks and should be used to their maximum advantage either in stratification or when comparing separate fields.

Multivariate EDA

After univariate and stratified exploration of the data, the natural progression is to move on to multivariate EDA. Whereas the univariate approach was primarily concerned with describing a single field, or in the case of stratification, sets of data within that single field,

Table 8.3 Correlation matrix (selected variables)

	Sale Price	*Lot Size*	*Home Size*	*Age*	*# Bedrooms*	*Condition*	*Quality*	*Neighborhood ID*
Sale Price	1							
Lot Size	0.306	1.000						
Home Size	0.675	0.358	1.000					
Age	0.039	0.263	−0.066	1.000				
# Bedrooms	0.321	0.213	0.611	0.013	1.000			
Condition	0.090	0.162	0.077	0.446	0.095	1.000		
Quality	0.608	0.070	0.643	−0.432	0.304	−0.152	1.000	
Neighborhood ID	0.067	0.024	0.043	0.118	0.032	0.062	−0.027	1

multivariate EDA is concerned mostly with relationships between fields. Like the univariate analyses, there are statistical and visual approaches. Stratification can also be done alongside multivariate EDA techniques.

A correlation coefficient is the simplest and most often used metric to test for the relationship between two fields. The correlation coefficient is a single value, ranging from -1 to 1, indicating how one field changes with changes in the other field. A value of 1 (-1) means that as one increases the other increases (decreases) in perfect unison. Values near 0 indicate the two fields change rather independent from another. Or, in other words, there appears to be very little relationship between the two.

Correlation coefficients are often calculated on many fields at the same time and the pair-wise comparisons are placed in a matrix. Correlations are usually calculated on continuous or discrete numerical values. For ordinal data there are a number of rank correlation procedures[1] available, while for categorical variables Cohen's Kappa offers one metric of relationship between fields.

Table 8.3 shows an example of a correlation matrix on a selected set of variables from Seattle sales data. Home size and quality show the greatest positive correlations with sale price (first column). Interestingly, there appears to be little relationship between age and sale price, likely due to the fact that most of the oldest homes are located in the most desirable downtown neighborhoods. However, there is a large negative relationship between home age and home quality, suggesting that many of the remaining older homes are not in as good of quality as the new ones.

The neighborhood ID field was added to show what can happen if categorical (levels stored as numbers) data is placed in a correlation matrix. If the categories are numbers, the calculations will work, but since the numbers have no order, the results are meaningless, as evidenced here by no real relationship with any of the remaining variables. Caution should be used when running correlation analyses to ensure that the variables used are continuous or discrete numerical/ordinal and not categorical ones stored as numeric values.

Visual approaches

Scatter plots are the most common visual method to examine the relationship between two fields. A scatter plot maps the value of each observation to a Cartesian plane (a two-dimensional X,Y space) where the X and the Y values are two of the fields in the dataset.

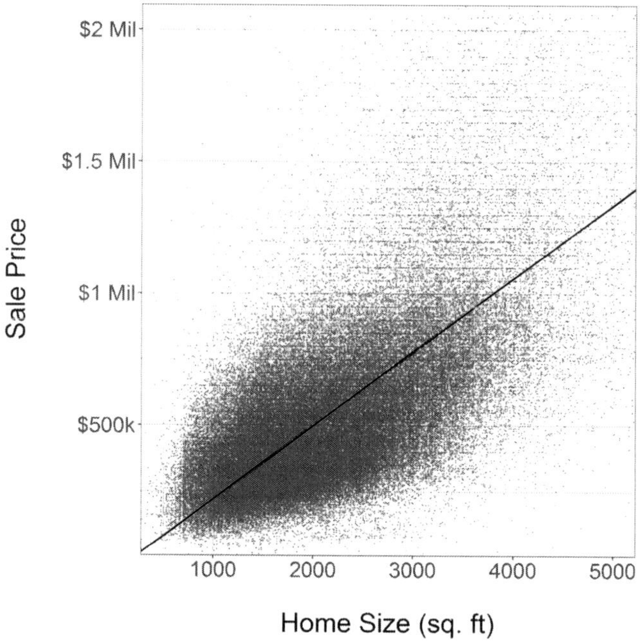

Figure 8.6 Scatterplot of sale price and home size

Figure 8.6 offers an example of a scatterplot showing the relationship between home size (x-axis) and sale price (y-axis) in the Seattle sales data. Not surprisingly, there is a strong positive relationship where larger homes are generally more expensive.

Scatter plots are also related to more advanced statistical measures. For example, a univariate linear regression model with the y-axis as the dependent variable and the x-axis as the independent variable will create a measure of the slope of the relationship on the scatter plot. In Figure 8.6, the solid line indicates the slope of the relationship as indicated by a simple linear regression model. Additionally, the R-squared statistic of the simple (univariate) linear regression model is the square of the correlation coefficient between the two variables (as shown in Table 8.3).

Like the correlation matrix, if there are several variables that need to be compared, scatterplots can be combined into a matrix to visualize a number of relationships at once. Figure 8.7 shows a scatterplot matrix of four variables from the sales data. From this approach a number of interesting trends and idiosyncrasies of the data are visible. Across the third row – Number of Bedrooms – there are a number of properties with 30 or more bedrooms, which are likely discordant values (data errors or outliers). Discordant values like the 30-plus bedrooms have the effect of skewing the axes and make the plots difficult to interpret. A researcher may want to temporarily hide those values or constrain the axes and re-plot. The techniques and process for removing these observations from the data will be discussed in the next chapter. Additionally, in the plots with building quality, the different levels are clearly visible as horizontal or vertical lines. This shows the difficulty (or danger) in using scatterplots with ordinal data having few levels.

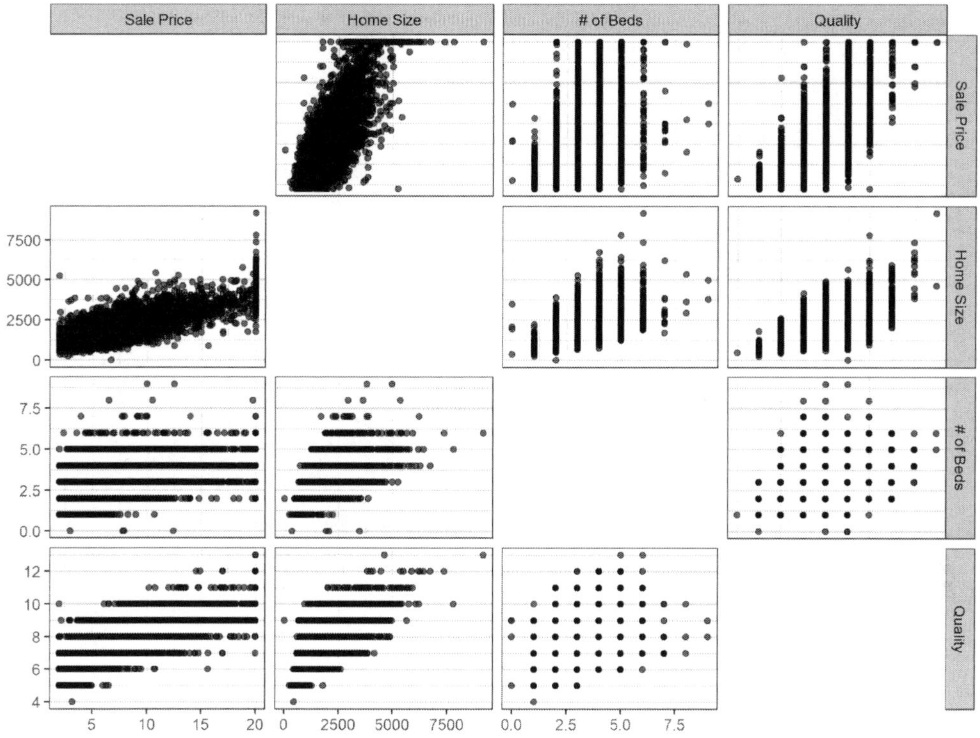

Figure 8.7 Scatterplot matrix (selected variables)

Mapping

For data with spatial coordinates, mapping can be an excellent visual approach to EDA. As most mapping involves at least the X and Y coordinates as fields to identify location, mapping can be considered a multivariate approach to EDA. The process of mapping can allow the researcher to both visualize the spatial extent of data as well as to determine if there are any spatial patterns in the field(s) analyzed. Mapping is a first step in a wide array of spatial analysis techniques used in EDA and discussed more fully below. The use of maps or other spatial plots can also highlight patterns that are not obvious from statistical measure or visual methods such as density plots or histograms.

Before mapping, the spatial identifiers and the spatial extent of those identifiers must be determined. To begin with, are the data observations geo-identified as points or as areas (polygons)?[2] If data are identified with points, a set of X (longitude) and Y (latitude) coordinates, then mapping can proceed. If the data are referenced by an area such as a state, county, or other administrative district, then the underlying geography of those areal extents must be gathered before mapping can occur.

Further, mapping may be of the individual observations themselves or of aggregated values at some coarser spatial level. While it is usually advisable to start mapping at the lowest level of disaggregation in the data and work up to larger aggregations, especially large datasets may benefit from aggregation at the onset.

Initial mapping will often first simply plot the location of each observation of the dataset, as seen in Figure 8.8a. This allows the researcher to understand the spatial extent of data as well as the density of observations in given areas and the propensity of the data to cluster together. A second, and more useful, approach is to then provide the observations with a value from one of the data fields to see if there is any spatial pattern to that field. Figure 8.8b shows the same observations as the left panel, only this time shaded by sale price. As can be seen, there is a definite pattern of clustering of high and low priced homes in the City of Seattle.

What point-based maps provide in granularity, they can lose in generalizability. Spatially aggregating point data into areal units, summarizing and then mapping can be useful as well. Figure 8.9 aggregates the sale price data by police beat and show the mean price in

Figure 8.8 Observations locations and sale prices

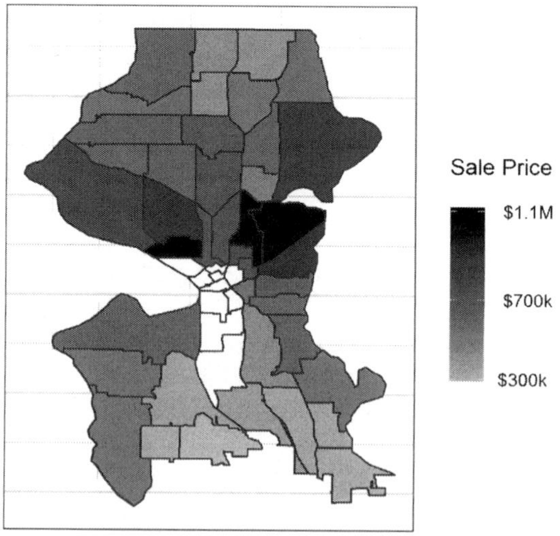

Figure 8.9 Mean sale price by police beat

each beat, indicated by shade. As with the point maps in Figure 8.8, we see significant levels of price clustering in the city. The beats with no shading (white) indicate empty beats in the center of the city that have no single family or townhome sales.

Spatial analysis

Spatial analysis is a broad category of analytical methods applied across a variety of disciplines, ranging from medicine and public health to economics and ecology. In the context of this book, a small component of spatial analysis is discussed as it is commonly used in exploring real estate data. As such, analysts consider the spatial analysis methods discussed here as an extension of mapping, whereby the spatial relationship between points or polygons (areal units) – both absolute and relative – are of interest.

One of the most common spatial analyses done during an EDA phase is the determination of the level of the spatial independence (or lack thereof) within a given field (Fortin & Dale, 2006). In other words, we test how often are high (low) values in a field located next to similarly high (low) values. There are several statistical measures that seek to measure the level of this relationship, known as spatial autocorrelation. The presence of spatial autocorrelation is important because it causes models to violate many of the assumptions on which standard econometric models rest. Finding spatial autocorrelation in the EDA phase may force the researcher to consider a different or more complex eventual analytical model than originally envisioned. Tests for spatial autocorrelation are also done during the modeling phase, as the lack of spatial independence in the model residual terms can signal problems of model bias and robustness.

One of the most common measures of spatial autocorrelation for a single variable is Moran's I (Fortin & Dale, 2006). Like the correlation coefficient, the Moran's I value is expressed on a scale from -1 to 1, where 1 indicates that high (low) values are surrounded by high (low) values, -1 the opposite, and 0 that no pattern exists. In practice, and especially in real estate research, negative spatial autocorrelation (values less than 0) is rarely found.

Table 8.4 shows examples of Moran's I calculations for five selected variables from the Seattle sales dataset. For sale prices, home size, and home quality, there are very high levels of spatial autocorrelation, indicated by Moran's I statistics above 0.35. Bedrooms show a lower (though still high) spatial relationship, whereas the date of sale field (just the date itself, not month or year) shows very little spatial autocorrelation, as expected. Future modeling exercises on these data will want to take into account the significant spatial patterns that clearly exist in the data.

Clustering methods are another common spatial EDA technique. The goal of clustering techniques is to determine if there are interesting subsets of data based on spatial location. For example, to test if there are clusters of very high-end homes in a dataset, we can run

Table 8.4 Moran's I measure of spatial autocorrelation: select variables

Variable	*Statistic*	*Z-value*	*P-value*
Sale Price	0.535	254.97	< 0.0001
Home Size	0.394	187.26	< 0.0001
Quality	0.438	208.57	< 0.0001
Bedrooms	0.173	82.47	< 0.0001
Date of Sale	0.006	2.77	0.002

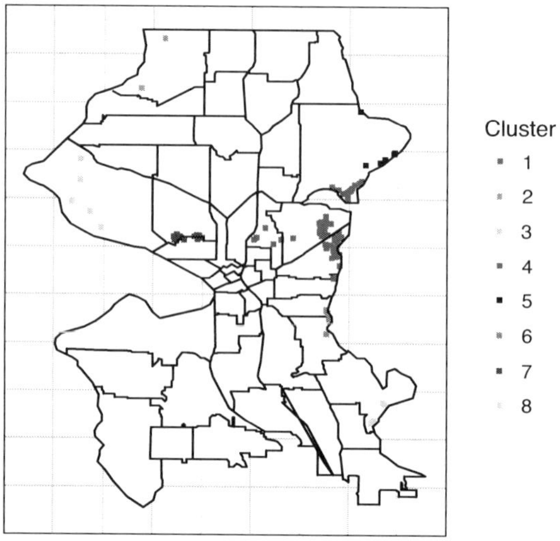

Figure 8.10 Clusters of high value homes

a clustering method on those homes to see if there are clusters and how they might be delineated.

Figure 8.10 shows the results of a hierarchical clustering method employed to all home sales greater than $3 million dollars in the Seattle data. From this analysis, we see that the algorithm has identified eight specific clusters of very high-end homes. This ranges from a large cluster of dozens of such homes in the center-right of the city (Madison Park neighborhood), to a few lone high-end homes in the far NW and SE parts of the city. Caution should be exercised with clustering methods as they are usual highly dependent on the parameters entered and/or can give different results over various iterations due to random starting points (*k*-means, for example).

Measures of spatial autocorrelation and spatial clustering methods are just two of the more popular spatial analytical techniques that can be used in the exploratory data analysis phase. More advanced methods are beyond the scope of this book and readers can look elsewhere for the theory behind, and examples of, these other methods.

Field substance

When exploring data it is important to consider the substantive characteristic or information point that each field represents. There are two dimensions of substance to be particularly aware of: (1) Behavioral vs. Contextual; and (2) Required vs. Optional.

First, a data field can be either contextual or behavioral, with the difference being defined by the research question and the particular level of observation of the data. Contextual data are those fields that provide the general situation surrounding the observation. Often, these are the time or the location of an observation. In other words, they set the context for the behavioral data. Contextual data are usually not the primary variables of interest. A researcher or analyst is often not concerned with the values of the contextual data themselves, only on how they provide context to the other fields. Examples may include the property's address or a contract/sale date.

Behavioral data, by contrast, are those fields that describe the characteristics of particular interest for a given observation. For example, in looking at office rents in a city, the time of contract, and location of the building would be contextual, but the price and terms of the deal would be behavioral. Depending on the situation, time and location may not always be contextual. If a dataset is measuring retail activity in a shopping mall, then time could be a behavioral characteristic (e.g., time of day when purchases are made). For the purposes of error or outlier identification, the limits to contextual data are often set by the study design itself; a certain area over a certain time and values outside of this range are either irrelevant or likely errors and are easily removed. Behavioral data, on the other hand, usually have no pre-set limits, and this is where most error or outlier identification action is needed.

To be clear, there is not always a perfect delineation between contextual and behavioral data, a priori. Instead, this split is determined by the research question and the type of analysis at hand. For the purposes of exploration, many multivariate techniques will pair a contextual field with a behavioral field, mapping being a prime example of this case. The distinction between the two will play a larger role in identifying and categorizing discordant values (Chapter 9), but the researcher should set out to identify which fields fall into which category before the data cleaning stage.

Second, a data field's substance can be required or optional. Required fields are those that cannot have a null value (not present or observed) in them, while optional fields are those that can have a null value. Contextual data will nearly always be required. As an example, in house price data, the sale date, sale location, and sale price are obvious required fields as a null value in these fields means that a sale (the level or entity of observation) did not occur. Similarly, if the data represent home sales, then the fields indicating home size and home age cannot be a null either, as all homes must have a size and an age (even if the age is 0). Conversely, if there are fields that indicate the presence of a pool or the size of a garage, a null value is possible in these fields since the lack of a pool or a garage can still signify a valid home transaction observation. For some fields, such as bedrooms or bathrooms, the difference may not be so clear and will depend on local conditions (studio condominiums or old structures with outhouse toilets). The distinction here is one that is entirely based on the domain of study; thus, there are few generic rules in its identification, and much of this relies on experience, local knowledge, and heuristic judgments.

Optional fields can be of any data type (binary, continuous, ordinal, categorical, or textual). Considering both the data type and its required or optional state is important for data exploration. Standard metrics such as central tendencies, distributions, correlations, density plots, etc., may provide misleading results for optional fields.

In the case of binary optional data (presence or absence), the mean simply represents the number of observations with presence of that characteristic being measured. For example, if the mean of a field indicating the presence of a pool is 0.18, that means that 18% of the observations have a pool. Binary optional fields are often used as a stratifying dimension in comparing the distribution of some other continuous, required field.

For continuous, optional data, the mean or median has no clear interpretation as observations with those characteristics are downweighed by those without. In these cases, the data may need to be converted to a binary (presence or absence of) or subsetted – mean taken only for those observations with the characteristics in the field – to gain any meaningful interpretation. For example, if a field contained the size of the garage, this could either be converted to a binary that shows what percentage of observations have a garage, or only those observations with garages could be analyzed to determine what the distribution of garage sizes are for those observations with garages. Ordinal and categorical optional variables may be used as stratifiers or subsetted depending on the number of categories or levels.

Within the data exploration phase, determining the substance of a field, contextual versus behavioral, and/or optional versus required, is critical to extracting the proper insights from the process. Making these determinations are also very important for the identification and treatment of discordant values (Chapter 9) and the statistical modeling (Chapters 10 and 11). The exploratory data analysis phase are recommended to fully determine the substance of all fields so that future analytical decisions can be most accurately and efficiently made.

Tools and documentation

Many of the simple statistical and visual methods for exploratory data analysis discussed above are common in standard point and click, GUI-based software, such as Excel and SPSS. Some mapping can even be done in Excel using plug-ins such as Power Map. However, these common products are often limited by the number of observations (rows) that can be analyzed as well as the number of fields in any given multivariate comparison. More advanced, code-based software, such as Stata, SAS, R, and Python have solid graphics packages that allow, with a bit of coding, efficient means to explore a multitude of relationships, subsets, and datasets during the EDA. Stratification and the use of small multiples (Tufte, 1990) can be especially painful in a point-and-click software environment, while being efficiently manufactured in code-based software alternatives.

For mapping purposes, Geographic Information System (GIS) software such as ArcGIS or its open-source equivalent (e.g. QGIS) is the easiest approach to EDA with spatial data. The spatial and mapping abilities of R and Python are rapidly improving and offer many of the same mapping options as the GIS product, albeit often with less interactivity.

Like the data management and data preparation steps discussed in previous chapters, exploratory data analysis represents a key component of the analytical workflow. However, unlike these prior steps, not all exploratory data analysis is typically recorded and documented as much of what is done in EDA is, by its very nature, exploratory and not worth repeating. Key insights from EDA that become part of the data cleaning process or those that support or validate subsequent analytical decisions (such as choice of estimator, measures of spatial autocorrelation, etc.) should be documented as they ultimately influence data provenance. In short, the researcher should perform copious amounts of EDA and then document all those tasks that ultimately influence the analysis in a measurable way. There is no perfectly objective process of EDA, so transparency is the best practice.

Notes

1 The two most common being Spearman's rank and Kendall's Tau.
2 Note that lines are the third form of possible spatial unit (along with points and polygons) but are somewhat uncommon in real estate data.

References

Behrens, J. T. (1997). Principles and procedures of exploratory data analysis. *Psychological Methods, 2*(2), 131.

Fortin, M.-J. & Dale, M. (2006). *Spatial Analysis: A Guide for Ecologists.* Cambridge, UK: Cambridge University Press.

Gelman, A. (2004). Exploratory data analysis for complex models. *Journal of Computational and Graphical Statistics, 13*(4), 755–779.

Peng, R. (2016). *Exploratory Data Analysis with R.* Victoria, BC, Canada: Leanpub.

Tufte, E. (1990). *Envisioning Information.* Cheshire, CT: Graphics Press.

Tukey, J. W. (1972). Some graphic and semigraphic displays. In T. A. Bancroft (Ed.), *Statistical Papers in Honor of George W. Snedecor* (pp. 293–316). Ames, IA: Iowa State University Press.

Tukey, J. W. (1977). *Exploratory Data Analysis.* New York: Addison-Wesley.

9 Data cleaning

Techniques to deal with outliers, errors, and missing data

It is often the case that the data management and preparation phases have not completely and totally filtered out observations for reasons of relevancy, or errors of structure, joining, or inconsistency. It is also highly likely that through the course of the exploratory data analysis (EDA) phase, potentially discordant values – errors of value or outliers – or missing data in some fields were identified. As a result, there is the need to apply a final set of discriminating criteria to the data to prepare it for the confirmatory analyses and modeling steps to follow. Table 9.1 shows a variety of rationales for removing observations from a data set.

This finalizing process is called "data cleaning." In this stage, the focus is primarily on the identification and treatment – removal or otherwise – of discordant values, such as errors of value and outliers, as well as observations with missing data. The resulting product of data cleaning is data that is ready for subsequent statistical tests, modeling, or visualization. This does not mean, however, that data is cleaned once and never again. The tests, models, or graphics that are generated afterwards may uncover issues that require yet more cleaning. As in all of the analytical workflow outlined in this book, data cleaning is often repeated multiple times as part of the iterative workflow process.

Errors vs. outliers

The difference between a data error and an outlier, by definition, is clear-cut. Data errors are incorrect data values that do not represent the actual, real world condition of the observation. Data errors stem from a number of possibilities including, but not limited to, key-punch issues, measurement problems, computer memory errors, data management and transmission issues (e.g., cell truncation), and fraud. Outliers, on the other hand, are correct data values, but ones that are not representative of the phenomenon of study and/or do not fit within the particular research question being addressed. As Hawkins (1980) puts it, "An outlier is an observation which deviates so much from the other observations as to arouse suspicions that it was generated by a different mechanism." In other words, while technically valid observations, they do not apply to the particular analysis at hand or are not representative of the underlying data generating process, findings, and conclusions.

While this difference may be very clear in terms of definition, determining if a particular value is one or the other in practice can be very difficult. For this reason, it is important to consider each to be a discordant value until it is properly identified as an error or an outlier. A number of the statistical techniques presented in this chapter will make that determination easier.

Table 9.1 Rationale for removing observations

Rationale	*When to Remove*	*Objectivity*
Error of Structure	Data Management	Objective
Relevancy	Data Filtering	Objective
Error of Join Matching	Data Filtering	Objective
Error of Obvious Inconsistency	Data Filtering	Mostly Objective*
Error of Value and Missing Data	EDA/Data Cleaning	Mostly Objective
Discordant Values (Outliers)	Data Cleaning	Somewhat Subjective

It is important to note that missing data can also be seen as a form of data error in which a measurement was not taken of a particular feature of a particular observation. In the case of missing data, however, the error is readily apparent due to the lack of a value, whereas in the case of many data errors their existence is not so obvious. Additionally, most statistical models, tests, and visualizations can function in the presence of data errors (errors of value or scale), whereas they may return error messages or produce strange results when given missing data. For this reason, missing data are often treated somewhat differently that suspected data errors.

Values properly identified as data errors can safely be removed or corrected. However, what if the researcher or analyst is not completely sure if a value is an error or just an outlier? What about known outliers? Can the observations with these values be "safely" removed from the analysis? The answer to that depends on the research question as well as the influence that these observations and/or values have on the final results and conclusions. In either case, as discussed in Chapter 4 on repeatable workflows, any value or observation removed must be logged and explained in the data provenance. In short, while removing an outlier may be a somewhat subjective decision, the reporting of it is a requirement. Chapter 10 discusses sensitivity testing and how it can be used to properly communicate the impact of data cleaning activities, as well as help researchers and analysts to decide which outliers to remove and which to keep.

Detecting discordant values (errors and outliers)

Many of the techniques and metrics at the center of the exploratory data analysis (EDA) process are also useful in detecting discordant values and determining if they are errors or outliers. As such, EDA and data cleaning can become an interwoven and iterative process where one leads to the other.

The simplest approach to detecting discordant values is to sort the data and manually examine the high and low values to determine if these values appear commensurate with the expected or possible values. Other slightly more advanced methods (discussed below) appeal to statistical measures or visual examination, such as asking if a value falls x number of standard deviations from the mean or if an observed value is at the tail end of a density plot.

Underlying these methods, but rarely discussed, are implicit assumptions about the distribution of the data in question. Simple sorting methods rely on the analyst's pre-conceived notion about what high and low values should be. In other words, the researcher may have an assumption about the appropriate distribution of the data. While domain expertise can certainly assist in making this a valid procedure, this assumption, nonetheless, does impact the data cleaning process. Likewise, identifying discordant values with metrics such as standard deviations and median absolute deviations (MADs) also implies an assumption about the distribution of the underlying data, most commonly a normal (Gaussian) distribution for continuous variables and a binomial or Chi-square distribution for discrete and categorical variables, respectively.

The assumption of normal or quasi-normal distributions is fairly standard in theoretical statistical methods; however, it may not always be appropriate in real world real estate situations. Many common data points, such as human heights and weights, are normally distributed because there are natural processes (often imbued with negative feedback loops) that control the data generating process. Economic data such as prices, costs, and returns are derived from data generating processes that more commonly create alternative distributions such as power law, exponential, etc. For example, there is a lower limit to prices, $0, but no actual upper limit. Most price data, therefore, are not normally distributed but exhibit heavy right hand tails.

As an example, Figure 9.1 shows the actual distribution of sales prices in Seattle over the 2011 to 2016 time period compared to a hypothetical normal distribution with the same mean and standard deviation. The most glaring aspect of the actual prices is the long right hand tail approaching the $10 million level. Even at greater than $2 million, there are a number of observations; if sales prices followed a normal distribution, there should be very few sales greater than $2 million. The hypothetical distribution also suggests sales prices under $0, which is not possible, given the nature of the phenomena being observed. Interestingly, the actual observations are more clustered (around $500k) then a normal distribution would suggest, which together with the long tail makes for a skewed, leptokurtic distribution.

Likewise, data that count discrete things that are often few in number, such as the number of bedrooms in a home or elevators in a building, usually follow one of the count data distributions, such as Poisson, Negative Binomial, or Zero Inflated Poisson. Network data,

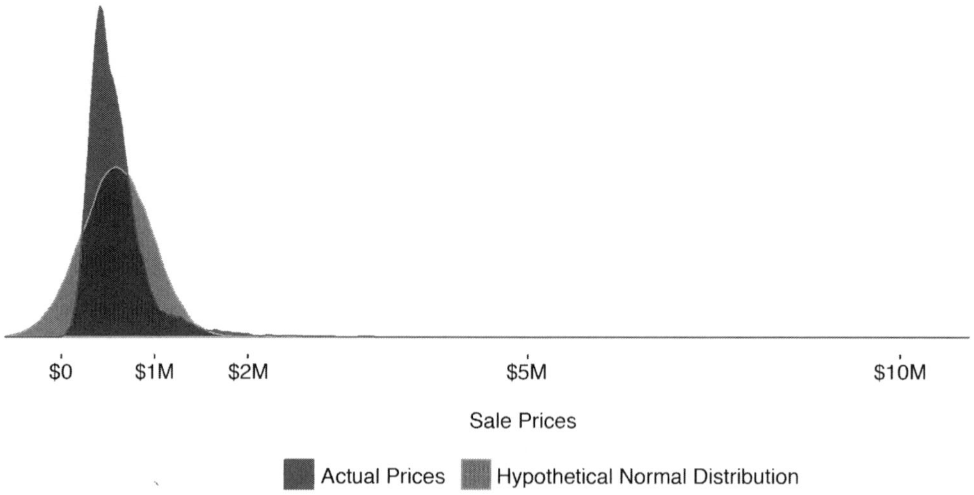

Figure 9.1 Distribution of sale prices vs. hypothetical normal

such as the number of Twitter followers, also tend to conform to non-normal distributions, as many people have few followers and a few people have many. In the era of big data there is no indication that the assumption of normality in the distribution of data should be one's *a priori* assumption.

As a result, before employing statistical tests to catch errors or identify outliers, the distribution of each of the data fields should be fully understood. As Chapter 8 notes, "getting to know" the data is a critical focus in Exploratory Data Analysis. Once the distribution of a particular data field or attribute is visualized, it should be compared with the theoretical or likely distribution of that particular feature of the data. Gross mismatches in actual and expected distribution can signal either some form of systematic error in the data collection process or that a researcher's initial assumptions and resulting research questions about the phenomena in question may need to be re-visited prior to further quantitative analysis.

Understanding the substance of the various data fields, as discussed in Chapter 8, is also a critical step in the data cleaning process. Discordant values in contextual fields should be treated differently than those in behavioral fields. Likewise, distributions and discordancy tests based on measures thereof require special attention for optional fields as compared to required fields. In short, the analyst or researcher must fully understand the nature of each data field prior to data cleaning.

Tests for discordant values can be divided between informal and formal. Informal tests rely on visual display and interpretation of the data and usually are accomplished through graphical output such as histograms, boxplots, or scatterplots. Formal tests, often referred to as discordancy tests, rely on statistical measures or other numeric evaluations to determine whether or not an individual value is an outlier. Often, potential outliers will be identified via an informal test first (often during exploratory data analysis) and then submitted to a formal discordancy test for further confirmation.

Single dimension tests (univariate)

Single dimension or univariate tests for discordant values are the most easily employed and, as a result, the most commonly used. These tests view an individual data field in isolation from all other information that is collected. For example, imagine a dataset that includes home price, living area, number of bathrooms, and the presence of a swimming pool. Applying a univariate discordancy test to the number of bathrooms field would only test for the appropriateness of a value within the believable range of values (the assumed distribution) for bathrooms. For example, a value of three would likely be acceptable under a univariate test, even if the observation was a 700-square foot studio apartment. In other words, three bathrooms is not an unbelievable value by itself, but it is if the property in question is a 700 square foot apartment.

The most common, univariate, informal tests are boxplots, density plots or histograms, described in more detail in Chapter 8 on exploratory data analysis. Typically, formal tests begin with an evaluation of standard or median absolute deviations. A standard deviation test computes the mean and standard deviation of given distribution of data and then applies a cutoff point beyond which, in terms of distance from the mean scaled by standard deviation (often called a Z-score or Z-statistic), an observation is considered an outlier. For instance, if a particular field has a mean of 10 and a standard deviation of 1, a 3 standard deviation test would label all values less than 7 and greater than 13 as likely discordant values. Often, a Z-score will be calculated for each value using the standard formula that standardizes the distance of each observation from the mean of the values in the field $Z_i = \frac{x_i - \bar{x}}{s}$, and all values greater than or less than a certain Z-score will be removed or labeled as discordant. The Z-score approach employs the same calculations as a standard deviation metric – the results are simply expressed on a different scale.

The median absolute deviation (MAD) criterion adopts a similar approach, except the median is used in place of the mean and deviations are calculated as an absolute difference instead of using the square root of the variance. Use of the median is more robust to biases in the mean due to the very outliers attempting to be identified and potentially eliminated. In a normally distributed field, the MAD will be lower than the standard deviation due to the squaring of the variance in the standard deviation. Multiplying the MAD by a factor of 1.483 creates the MAD_e, which is similar in scale to the standard deviation of a given data vector.

Tukey (1977) created a more nuanced metric that utilizes the interquartile range to create "fences" in the distribution of the data. The interquartile range (IQR) is the distance between the 25th percentile (q1) and the 75th percentile (q3). Tukey's method creates inner fences at q1 – (1.5 * IQR) and q3 + (1.5 * IQR) and outer fences at q1 – (3.5 * IQR) and q3 + (3.5 * IQR). Observations between inner and outer fences were called "outside" and those outside of the "outer fences" as "far out." While this method may seem mildly subjective, the fact that it does not rely on the mean or the standard deviation means that it may be more robust in the presence of non-normally distributed data.

Additionally, more complex discordancy tests for univariate errors or outliers may also be employed. Grubb's Test, the Tietjen-Moor Test, Peirce's Criterion, and Chavenet's Criterion are some examples of these tests.

Multiple dimension tests (multivariate)

Single dimension or univariate tests may fail to reveal discordant values in cases such as the three bathroom, 700-square foot studio condominium noted above. For this reason, testing for outlying values across multiple dimensions is often necessary. Like the univariate tests, both informal and formal multivariate tests are available.

Scatterplots present the most common informal test to identify errors or outliers. By plotting one dimension or field of data against another, researchers are able to discern (visually) if any of the values do not appear to follow the trend in the relationship between the two fields (if there is a trend). From a simple two-dimensional scatterplot, three types of discordant values can arise:

1 Conditional discordant: An extreme value in one of the dimensions, but only when related to the other dimension
2 Leverage case: Extreme in both dimensions but fits overall trend of the relationship
3 Contaminant: Extreme in both dimensions, not fitting trend

Figure 9.2a shows an example of a conditional discordant where both the home size (measured in square feet) and the home price (measured in dollars), on their own, are reasonable values; however, conditional on the home size the sale price appears discordant (or vice versa). Figure 9.2b shows an example of a leverage point. In this example, both the home size and the sale price appear very high; however, the relationship between the two follows the pattern of the other observations. Finally, Figure 9.2c shows an example where

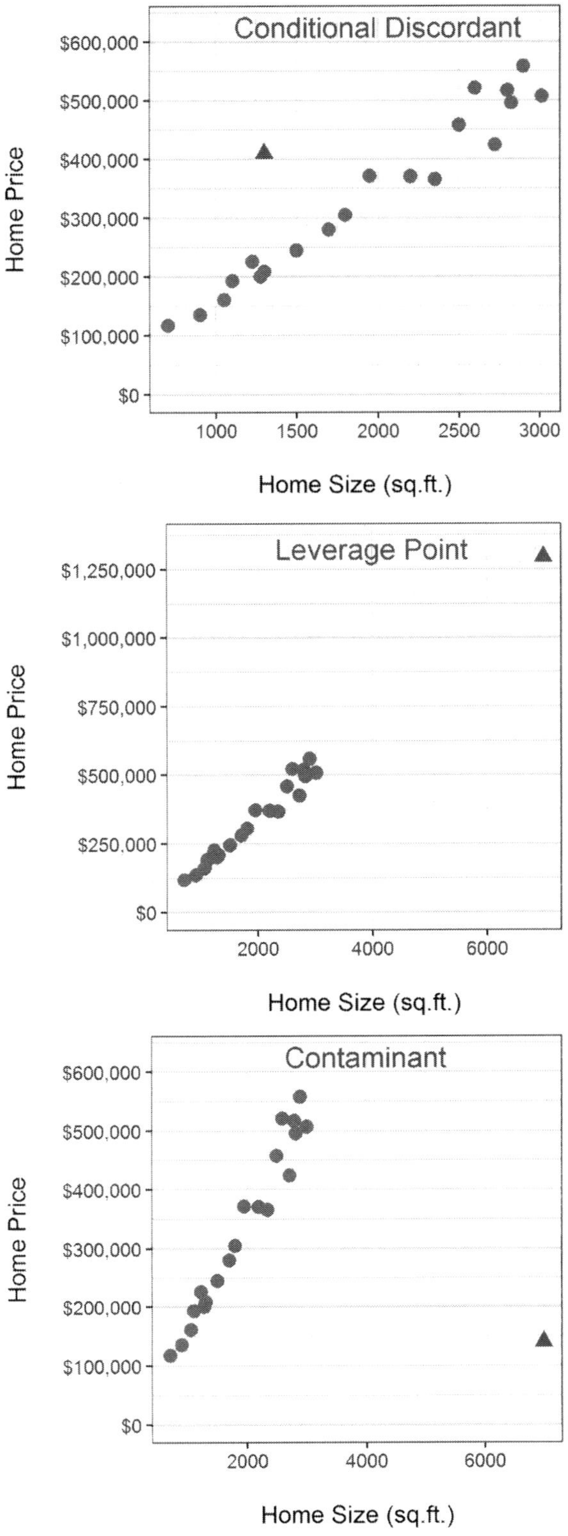

Figure 9.2 Examples of multivariate outliers

the home size is much higher than the rest of the data and the sale price is relatively very low. In short, the relationship between the two variables also appears out of line with the other observations. This is called a contaminant case, which occurs when an observation appears to be from a different distribution of data or data generating process relative to the rest of the observations. In the case of real estate data, perhaps the contaminant observation comes from another type of property.

In helping to decide between errors and outliers, conditional discordants offer the highest probability of a data error, while the leverage point in Figure 9.2b is likely a legitimate, but outlying value, and the contaminant in Figure 9.2c is probably a mislabeled industrial or warehouse property, one that should have been removed in the data preparation stage due to relevancy (were it properly labeled as such).

It is very important to note that not all conditional discordants are errors, nor are all leverage cases or contaminants valid observations. These are simply designations that can assist the researcher in deciding between errors and outliers. The ease at which discordant values are given one of these three labels, and by which errors may be distinguished from outliers, are driven by the distributions and the strength of the relationships that can be found between the fields in question. With very strong relationships over a well-defined distribution, it is easier to decide between conditional discordants, leverage points, and contaminants when distributions are unclear and relationships very weak.

Formal multivariate tests

There are formal tests for outliers within multivariate situations as well. Two of the more commonly used processes are Mahalanobis Distance and Partitioning Around Mediods. The Mahalanobis Distance method functions much like the Standard or Median Absolute Deviation method in one dimension, with the exception that the covariance between the multiple dimensions is considered in the calculation along with the mean and standard deviation. Again, like the standard deviation method, the Mahalanobis Distance method may be subject to influence from very large value outliers unless a robust estimation procedure is used. The indicative criterion is the (statistical) distance of the observation from the central tendency of the entire distribution.

Partitioning Around Mediods (PAM) is one of several clustering algorithms that also can be used to identify outliers. PAM iteratively sorts observations into multivariate clusters. Observations that fall outside of these clusters or a small group of observations in a small cluster distant from the larger cluster may indicate outlying observations. The general idea is that normal observations will be located in neighborhoods (not necessarily spatial in nature) with a high density of other observations, whereas outliers will be rather lonely. Both the PAM and Mahalanobis Distance methods can handle more than two dimensions at a time. Finding outliers in three or more dimensions can be difficult with informal methods such as scatterplots.

To handle the difficulty posed by outliers in multiple dimensions, Hadi (1992) proposed a method to handle multiple outliers in multivariate space. Related to what PAM tries to achieve, Hadi's method approximates the minimum volume ellipsoid (MVE) using an algorithm that is computationally more efficient and is not subject to the problems that typically plague attempts to approximate MVE, including *a priori* determination of the number of subsamples, subsamples with singularity of the covariance matrix, and covariance matrices from some subsamples that have zero determinants and thus ellipsoids with zero volumes.

Like the multivariate informal methods above, clustering- or distance-based methods will work best when relationships are strong and distributions well defined. In their basic format, most clustering or distance methods will only identify discordant or groups of discordant values – but they will not suggest their form of discordancy (contextual, leverage, or contaminant). Further investigation into the location of the clusters or the direction of the distance-based deviances may be necessary to classify the type of discordancy.

Contextual discordants

Datasets with contextual data fields allow for additional, and often specialized, discordancy tests. These methods can help to identify contextual discordants, many of which are data errors. Like the above, these tests can be both formal and informal. Informal tests for contextual discordants generally take the form of scatterplots or maps. When looking for contextual discordants in temporal or time series data, a scatterplot is typically employed where the x-axis is the time axis (contextual data field) and the y-axis is the observed, behavioral field in the data.

In Figure 9.3, the monthly median home value trends are shown for the city of Seattle sales data from early 2011 to the end of 2016. The observation in July 2012 (deliberately edited by the authors) has a value close to $550,000, certainly within the realm of possible values for Seattle over this time period. However, given the time or the context of that data point, the value of $550,000 is unrealistic in this case as the surrounding values in time are closer to $430,000. Here the contextual data (time) has pointed to a discordant value in the behavioral data (value).

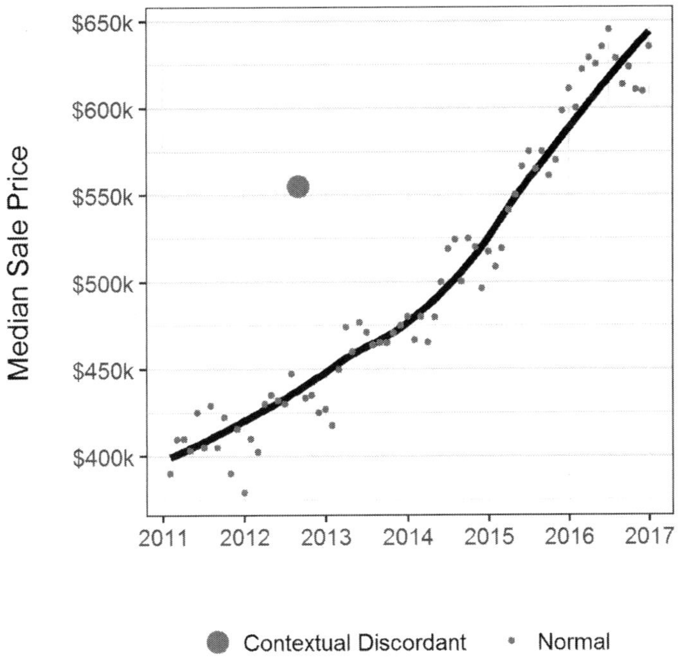

Figure 9.3 Examples of contextual discordant

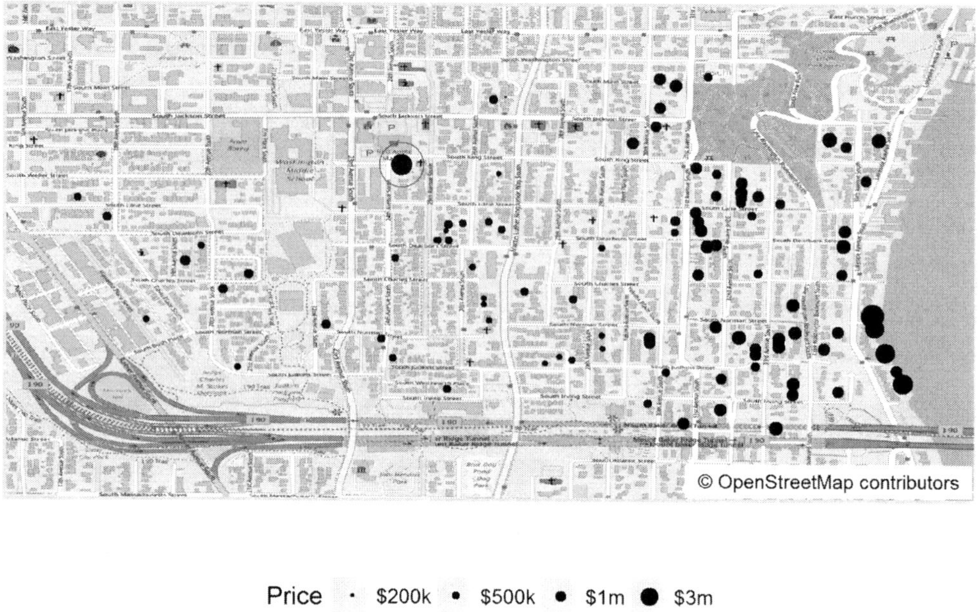

Figure 9.4 Example of a spatial contextual discordant

When geographic space is a contextual variable, mapping is a common form of informal discordancy testing. Much like the scatterplot in Figure 9.3, a mapping approach relies on visual identification of a value that does not fit the context of its immediate surrounding. Figure 9.4 shows an example of a contextual discordant based on the location of an observation. The points on the map are sized by their sale price, with larger sizes equal to larger prices. The high price sales along the right hand side of the map are waterfront and the high prices can be explained by this amenity. As one moves west, water access and water views diminish and home prices drop, with the exception of the one highlighted sale in the upper middle of the map. Overall, looking at sale prices in this area, the price is not discordant, but given its spatial context it certainly is discordant. It is a manufactured data point for the sake of this example.

Much like the formal tests of global spatial autocorrelation that were discussed in the Exploratory Data Analysis chapter, formal tests for identifying contextual *spatial* discordants exist as well. Two of the most common tests are Local Moran's I and Local Geary's C. Although the exact calculation differs, both tests compare each observation to its local (spatially proximate) neighborhood in order to determine if its observed value is like or unlike the values around it. The statistics – Moran's I and Geary's C – are expressed in standardized values that are akin to Z-scores. A statistical significance test is also available such that one is able to test how likely the variation between an individual observation and its neighbors is due to random chance or is a result of it being an actual outlying observation.

Figure 9.5 shows the results of a local Moran's I test[1] on the observations in this neighborhood. The square points show observations with statistically significant and positive local spatial autocorrelation, a phenomenon common in real estate. The lone triangle point is

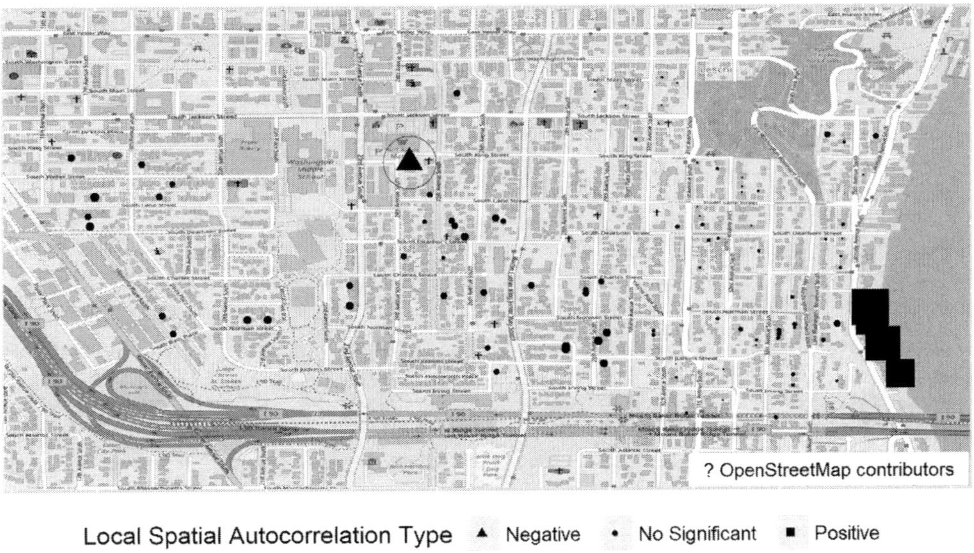

Figure 9.5 Local spatial autocorrelation of sales prices

the suspected contextually discordant value and the only observation to show statistically significant and negative local spatial autocorrelation. Given this, it is likely that this observation is a contaminant (special use property or similar).

The issue of global versus local outliers is not constrained to data with spatial context. The "local-ness" of an outlier can refer to its proximity in any dimension. Further, the importance of local outliers can be magnified or dampened based on the ultimate modeling strategy of the analysis. If the research plan calls for localized analyses or clustering algorithms, local outliers may introduce more bias than if the final empirical models will approach the problem from a global perspective only.

The methods and interpretation of the tests for conditional and for contextual discordants are often identical. The key difference between the two is that in the conditional case, the two (or more) dimensions that are being analyzed are all behavioral data; in the contextual case at least one of the dimensions is contextual data. Additionally, the strength of the conclusion that can be reached may favor contextual discordancy tests as they usually provide more certainty of observations being errors than conditional discordants due to the stronger relationship in contextual data over space and time.

Results-based outliers

In the above cases, we have discussed analyzing the data for outliers prior to running any models or other analytical tests. Many of the methods above can also be used to examine model results or other test statistics for the presence of outliers. In this case the researcher is not necessarily looking for observations that are outliers in their raw value, but rather are observations that behave discordantly within the applied statistical model or test.

Commonly the residuals or errors of a regression model, such as a hedonic price model, are submitted to univariate discordancy tests. Since one key assumption of linear regression models is spherical, normally distributed error terms with a mean of 0, a Q-Q plot is often used. A Q-Q plot displays the quantiles of one data field against the quantiles of another. In the case of model residual analysis, the quantiles of the actual residuals are plotted against a hypothetical set of residuals that follow a normal distribution with mean 0.

For data with spatial attributes, a Local Moran's I or Local Geary's C test can be used to determine if the residuals for certain observations do not match the overall pattern in their area. Models using spatial data often first employ Global Moran's I or Global Geary's C to test for spatial autocorrelation in the residuals of the models. Once model specifications have been adapted to control for global spatial autocorrelation, discordancy testing with the corresponding local statistics can begin.

Caution should be exercised when testing for discordancy in model-based outcomes such as predicted (fitted) values or residuals. Incorrect or insufficient model specification can lead to large residuals and/or spurious fitted values. In such a case, observations that appear to be errors or outliers based on the test noted above may be simply the artifact of an incorrect model. Models should be rigorously tested and diagnosed for common modeling errors and issues prior to using model outcomes to identify discordant values in the data.

Additional considerations

There are a number of additional items to consider when employing discordant identification methods. The first is whether the researcher is attempting to find discordant values one at time (single) or multiple at once. Some tests, such as standard deviation methods, can prove difficult to identify multiple discordant values at once, especially in the presence of extreme values. There are two basic mechanisms or effects by which identification of multiple errors or outliers can be hindered: masking and swamping.

Masking occurs when one error or outlier's extreme value masks or prevents the ability of a test to identify other discordant values. Imagine a situation where an analyst is using a 3 standard deviation limit to determine discordant values. The dataset is rather small and one value is 10 times higher than the mean of the others. As a result, the standard deviation is greatly inflated and only the one extreme value is labeled discordant, masking the potential to identify other outlying, but not as extreme, values. Discordant identification methods that only use a single step often suffer from masking effects.

The swamping effect produces a somewhat opposite effect. In tests where the analyst specifies, *a priori*, the number of discordant values to identify, specifying too many errors or outliers can cause the identification of one observation as an outlier to lead to the identification of a second observation as a discordant value as well, the latter which would not be so except that the first observation is labeled as such. This effect is also common in sequential discordant labeling and removal, whereby eliminating the most extreme values reduces the standard deviation drastically and causes more observations to appear as errors or outliers. As the iterations increase, more and more of the less outlying observations appear to be errors or outliers due to the shrinking standard deviation (or other test statistic) being used.

Masking and swamping effects can be curtailed through the use of tests that are robust to the presence of extreme values. The simplest robust metric is to use medians instead of means as they are less prone to bias from values at the tail end of various distributions.

For univariate tests, this may mean replacing standard deviation metrics with Median Absolute Deviation (MAD). In the multivariate situation, many of the tests mentioned may have robust options depending on the analyst's statistical software package. Informal tests themselves may be less susceptible to extreme values; however, software packages that automatically scale the x- and y-axis on plots may automatically produce a swamping or masking type effect. Transforming axes into logs or creating custom plots can help analysts avoid being fooled by the scale of informal discordancy test plots.

Handling errors and outliers

Once discordant values are identified what should be done with them? The first step is to confirm whether or not it is an actual outlier or a data error. In most cases with large datasets, observations with data errors can be removed so long as the errors are few in number and random across the data. If there are many errors or if they are systematic, then the errant data points should be treated as missing data and the advice in the following section on missing data should be followed. In the event that an analyst is working with a small dataset and the loss of an observation may be detrimental to the validity of the analysis, then an attempt to correct the errant observation can be made. When dealing with big data sets it is not always practical to correct individual observations, so business rules or data filters that can be applied to the entire data set are preferred.

After the known data errors are reconciled, a decision on how to treat the remaining discordant values – the outliers (known and suspected) – can be made. Essentially, three choices are available: (1) Remove; (2) Accommodate; (3) Label and Test.

Removal of outliers is the most straightforward and easiest choice of the three. However, it is often the most difficult to defend to an audience as one has to prove that the observation removal was not done to generate a desired outcome. Outlier removal can and should occur when the discordant observations do not fit the research question and the assumed data generating process of the data. The removal of confirmed contaminant data points (see Figure 9.2) is a good example of when observation removal is the preferred choice.

Outliers can also be accommodated in the statistical methods and models that make up the empirical analysis. Most statistical models and techniques now have robust options for their estimation. These robust estimators minimize the impact or leverage of outlying observations usually through some iterative weighting mechanism. Robust regression is one such technique that down-weights the impact of outliers on the "line of best fit."

The third option is to label the outliers in the data and then test a series of models or provide estimates with and without the outliers to determine their impact on the results. This process is often referred to as sensitivity analysis and provides a very transparent view of the process used and the variation in results due to potential outliers. The outlier labels do not have to be binary ones and zeros. Outliers can also be scored based on their atypicality relative to the other observations and a number of sensitivity tests run to see how different cut-off points for outlier considerations may impact the results. While sensitivity tests and multiple model runs was once a considerable extra work load for an analyst, those following reproducible research tenets (coded and repeatable analyses) will find the Label and Test method to be very easy to accommodate as well as the most transparent way in which to approach the presence of outliers. Note, however, that the Label and Test process may lead to (1) the removal of some of the most extreme values; or (2) the use of robust statistics if the analyst finds that the results are sensitive to certain outliers and a reasonable and supportable argument can be made for the removal or down-weighting of those observations.

Missing data

Most raw data encountered by analysts in the real estate industry is not primary data collected on their own; most data encountered in the real estate industry is secondary data. As such, the analyst has little control over the initial collection and management of the data by the originator. Missing data values are common. Missing data that was simply not collected can be considered data measurement errors, as there is a lack of measurement. If the missing data is the result of storage and management issues, it is an error of another type.

In the absence of a data value, the initial task is to determine if the data value is a necessary value or simply the lack of an entered value. For required fields (see Chapter 8), any missing value can safely be assumed to be missing data. On the other hand, for optional fields the analyst must confirm if a null or missing value is actually missing data or merely the lack of the feature represented by the field. While failure to correctly indicate the lack of a feature in the data is poor data collection and management practice, it is also one that is all too common in secondary real estate data. To date there is little in the way of standardization across real estate data collection agencies and the primary data inputters are often far removed from the analysis process and have little training on proper data management techniques. As a result of this situation, a researcher should always use caution when interpreting missing values in real estate data.

If the empty values can be confirmed as missing data, the type or structure of the "missing-ness" – the pattern of the missing data – must be determined (Van Buuren, Boshuizen, & Knook, 1999; King, Honaker, Joseph, & Scheve, 2001). Missing data can be of one of two types: random or not random. Missing at random (MAR) data is the less serious of the two. In a sufficiently large dataset, especially one entered by hand, it possible that some values have been randomly omitted due to key-punch errors, computer memory errors, or other problems in the data integration process. Since a large portion of real estate data is hand entered and combined from multiple sources, missing at random data is always a possibility.

However, data that is not missing at random (NMAR) is much more common and, unfortunately, more problematic as well. Data that is NMAR means that the probability of a given data value being missing depends on or is related to the missing value itself. In other words, if certain information is harder to collect, for example, due to its actual data value, then failure to collect it causes data to be NMAR. For example, if the age of building is difficult to determine for structures built prior to 1950, and for those properties the age field is left blank, then this data is NMAR because the likelihood of it being missing is related to its value. NMAR issues also often arise when combining data from different sources, especially when one source is collected in the field. In this case, if the observations from each source are not perfectly randomly distributed among the entire distribution of all observations (i.e., data collection by hand is often done in geographic clusters for efficiency reasons), then the data is NMAR.

Once the type of missing-ness is determined, the analyst then decides on how to remedy the situation. There are essentially two options available to the analyst: (1) removal of observations with missing data; or (2) imputation or correction of the missing data. The removal of any observations with missing data leaves behind so-called complete cases (CC) or complete observations. The complete case method is most applicable in instances where there are few missing data values relative to the overall size of the data set and/or where the missing values are determined to be MAR. CC analysis in the situation where there are many missing values, and where those values are NMAR, will likely create a remainder sample that is different from the original. The severity of this potential bias in sample constitution is determined by both the number of removed cases and the importance of the missing values to explain the phenomenon in question (Little, 1992). In other words, missing dependent variable or critical independent behavioral or contextual data is more damaging to the analysis than missing values in a data field that is unlikely to be used in the final model specification. Again, sensitivity analyses can be used here to determine the impact of the reduced sample resulting from CC on the final results. The potential for and magnitude of these biases should be discussed fully by the researcher and avoided if possible.

If CC analysis proves infeasible, the missing data can be imputed, or in a minority of situations corrected with additional data collection. The most common imputation method uses a predictive model where the missing data field is the dependent variable and the complete case observations are used as independent variables. Some form of regression or other statistical model is then used to predict the missing values. As with any predictive model, the accuracy of the imputed values will depend on the strength of the relationships between the dependent and the independent variables. If the data in question has spatial attributes and there is found to be spatial autocorrelation in the known values of the field with missing data, then the spatial dependence of the phenomenon may also be used to assist in imputing the missing values. Overall, the imputation of missing data is multi-faceted technique, and for greater details on this process, the reader is directed to see Pigott (2001) or Little and Rubin (1987).

Conclusion

In this chapter, we have discussed the data cleaning process. We consider data cleaning to be a finalizing and analytical process whereby outliers and errors of value are identified and a strategy is developed on how to handle these discordant values.

The data cleaning process is a natural extension of exploratory data analysis and the two may often operate in an iterative fashion. For both EDA and data cleaning there are formal (statistical) and informal (visual) univariate and multivariate techniques that are used. Many of the same EDA metrics are used in the data cleaning step to identify discordant values.

Determining the difference between errors of value and outliers is not easy, nor even necessary in all cases. We present a breakdown of discordant values into conditional discordants, leverage cases, contaminants, and contextual discordants. Classifying the discordant values in your data set into these categories can help determine if the value is an error or an outlier.

Errors, properly identified, can be either removed or corrected. For discordant values determined to be outliers, there are three choices: remove, accommodate with robust models, or label and perform sensitivity tests. The research question at hand, the level of discordancy, and the quality of the data will all help to determine which choice or combination of choices is employed to deal with the outlying values.

Finally, missing data can also be considered a form of data error. Missing data can be dealt with either by complete case analysis or by imputing the missing values. Regardless of the method employed for discordant values and/or missing data, documentation of the data provenance and the rationale for removing any observation from the data is necessary.

Note

1 For more detail on the spatial weights matrix used in the local Moran's I test, see the accompanying code to this book located at www.github.com/REAIABook/REAIABook/.

References

Hadi, A. S. (1992). Identifying multiple outliers in multivariate data. *Journal of the Royal Statistical Society, Series B (Methodological), 54*(3), 761–771.

Hawkins, D. M. (1980). *Identification of outliers* (Vol. 11). London: Chapman and Hall.

King, G., Honaker, J., Joseph, A., & Scheve, K. (2001). Analyzing incomplete political science data: An alternative algorithm for multiple imputation. *American Political Science Association, 95*, 49–69.

Little, R. (1992). Regression with missing X's: A review. *Journal of the American Statistical Association, 87*(420), 1227–1237.

Little, R. & Rubin, D. B. (1987). *Statistical Analysis with Missing Data.* New York: John Wiley and Sons.

Pigott, T. D. (2001). A review of methods for missing data. *Educational Research and Evaluation*, 7(4), 353–383.

Tukey, J. W. (1977). *Exploratory Data Analysis.* New York: Addison-Wesley.

Van Buuren, S., Boshuizen, H., & Knook, D. (1999). Multiple imputation of missing blood pressure covariates in survival analysis. *Statistics in Medicine, 18*(6), 681–694.

10 Modeling real estate prices

Models that estimate prices and/or rents are some of the most common and complex real estate analyses employed in both industry and academic research. At the core of most pricing models are multiple regression analyses. These econometric analyses are also referred to as hedonic price models in the context of real estate pricing studies.

Hedonic price models, and their associated derivative models, are a particularly useful approach to analyzing real estate markets because they can be used to predict transactional values (prices or rents) as well as estimate the marginal (additional) impact of various property characteristics on the transactional values. These two approaches are often referred to as engineering (prediction) versus science (impact evaluation). Although a single hedonic price model specification can simultaneously do both engineering and science, some differences in approach can occur based on the desired outcome, be it engineering or science related.

Models built solely for the purpose of price or rent prediction will focus heavily on reducing standard errors and on running cross-validation tests to determine out-of-sample accuracy. Changes to model specification will be normally be focused on lowering error rates. Conversely, if the intention of a model is to determine the most accurate estimate of the impact of a given factor on price, then diagnostic tests concerning multicollinearity, heteroskedasticity, and spatial autocorrelation will likely be the focus of the analyst's time. Considerable effort will be put forth to ensure that the coefficients results from the model estimation are unbiased and efficient. In short, the goals of a project will drive the modeling process.

Overview of pricing models

Hedonic pricing theory is based on the assumption that the total value of a property is the sum of the value of its individual components, therefore, there is an implied value for each of the various property characteristics. For example, each bedroom has a specific marginal (or contributory) value, as does each bathroom. This value is not, however, limited to the structural characteristics of the property. Location factors such as proximity to transportation, proximity to employment centers, and/or proximity to shopping opportunities are major value drivers, too, along with the demographic features of the neighborhood. Intangible assets, such as the right to lease, develop, and transfer ownership interests either in part or in whole are also part of the value equation.

Building upon the work of Lancaster (1966), hedonic theory was first presented in 1974 by Sherwin Rosen in the *Journal of Political Economy*. Rosen posited that some commodities are *differentiated goods*; goods that are comprised of a set of characteristics that differentiate them from other, similar goods in the same market. Housing fits this definition quite nicely – each property includes a unique set of characteristics (bedrooms, bathrooms, quality conditions, etc.) that comprise, in the mind of consumers, a single composite good (the property, which includes the house and/or other physical structures that are a part of it). Cereal and automobiles are other examples of differentiated goods.

Suffice it to say that the number of peer-reviewed studies that use the hedonic model to extract the value of different housing characteristics far exceeds the space allotted for this chapter, even for this book. Des Rosiers and Theriault (2008) attribute the effectiveness of hedonic price models to the use of multiple regression analyses, which produces the most probable price (or market value) that can be expected based on the set of residential properties analyzed. A few examples of articles include studies of the pricing of local amenities such as golf courses, ocean views, parks, transit-oriented development, and shopping centers. Research on the value of disamenities, such as proximity to landfills, oil spills, leaking underground storage tanks (LUSTs), and crime are also widely available. There are so many articles on hedonic pricing that we leave it to the reader to investigate academic and practitioner sources for other examples (see *Journal of Real Estate Research, Land Economics, Real Estate Economics, Journal of Real Estate Finance and Economics, The Appraisal Journal, Journal of Property Investment and Finance*, the *Journal of Property Research*, and others).

Many uses for pricing models exist outside of the confines of academic studies. When used in a valuation context, such models are often referred to as mass appraisal models. The Appraisal Institute describes mass appraisal as the process of valuing a universe of properties as of a given date using standard methodology, employing common data, and allowing for statistical testing. Examples of mass appraisal tools include CAMA (computer assisted mass appraisal) models and AVMs (automated valuation models). In both of these examples, hedonic pricing plays an integral role in estimating market value. Mass appraisal techniques are regularly used in markets with a reasonably large number of properties that transact on a regular basis (the "mass" in mass appraisal). This technique is not appropriate for small or stagnant markets with few transactions.

Pricing models may also be used to generate house price indices, a feature discussed more fully below under longitudinal models. Lending risk evaluation has increasingly relied on statistical pricing models to value collateral as well as to generate portfolio level risk assessments. Finally, tax valuers (assessors) and/or property valuers (appraisers) may use the coefficient estimates – the marginal impacts – derived from the model(s) to justify adjustment made to comparable sales in more traditional valuation approaches.

Hedonic model basics

When applied to housing, the basic regression model includes a matrix of structural (S), location (L), and time (T) characteristics that are used to estimate prices (P). S, L and T are collectively known as independent variables, i.e., those variables that are studied to see if they affect the dependent variable, which is in this case P. The error term or residual (ε) captures any unexplained variation in prices that are not explicitly reflected in the model. Mathematically, the basic hedonic regression model for housing is expressed as:

$$P_i = a + bS_i + cL_i + dT_i + \varepsilon_i$$

Prices are commonly regressed against structural characteristics such as the number of bedrooms, number of bathrooms, size (square feet or square meters), condition, and age (either actual or effective age of the structure). While the list of preferred location and neighborhood variables regularly priced in the models is less consistent than the structural characteristics, common examples include proximity to public schools, proximity to highway access, distance to the central business district (CBD), and certain neighborhood demographic characteristics. Time variables refer to the date of sale, which often serves as a proxy for broader market conditions, capturing interest rate movements, changes to policies as well as property appreciation/depreciation over the long-term.

Many features in real estate markets, both dependent and independent variables, do not follow a normal distribution. As a result, variables are frequently transformed in house price model specifications. The most common is a log transformation of the dependent variable (price or rent in the case of commercial/investment property). Other independent variables may also undergo log or polynomial transformations if the distribution of the variable is non-normal (log) or if the relationship between the independent and the dependent variable is non-linear (polynomial). Additionally, since non-linearities are common among discrete, ordinal, and categorical variables, they are often transformed into a series of binary (dummy) variables to estimate the impact of each level or category on price (rent). Overall, it is important to note that the appropriate model specifications for each research project will be related to the question the research seeks to address and the market from which conclusions are to be drawn; thus the variables used may differ somewhat from the examples presented here.

Time

Real estate markets are highly dynamic, thus properly measuring and controlling for market trends is an important component of model specification. The most basic form of analysis used in real estate research is arguably the cross-sectional analysis. This method is used to evaluate data from a population (or a representative sample of the population) at a specific point in time or over a relatively short period; in real estate this generally means looking at the distribution of values during a particular time span without regard to the value of properties in the population prior to that. A study that compares office property rents or cap rates in multiple cities in 2007 would be considered a cross-sectional analysis, for example. However, even within a cross-sectional analysis covering a single year, there may be quarter-to-quarter or month-to-month movement in prices, rents, or rates that can and should be accounted for.

In a cross-sectional model covering short time periods, the impact of market changes over time are generally accounted for in one of three ways: (1) linear trends; (2) binary or dummy variable per period; and (3) polynomial or other advanced transformations. A linear treatment of time involves entering a measure of time – the count of quarters or month from the beginning of the study – as a continuous independent variable in the model specification. This necessary assumption implies that the market has changed smoothly in the same direction over the duration of the study (e.g., a consistent appreciation or depreciation rate). A more common treatment is the transformation of time into a series of period-specific binary or dummy variables. For example, a study spanning two years might create eight binary variables, one each for the eight quarters occurring in study period. The coefficients of these variables then represent the change in the market from the reference period (usually the first quarter) to each of the remaining seven quarters. Finally, more complex treatments of time such as polynomial or Fourier (McMillen & Dombrow, 2001) transformations allow for parametric treatment of market changes without strict linear confinements. In other words, cyclical changes in market movements can be modeled directly.

In cross-sectional models, market changes are usually a factor that needs to be controlled in order to properly estimate predictive values or to measure the impact of other coefficients on prices or rents. Conversely, models can be specified where the changes in time are the variable of interest. Such models can be considered time-series or longitudinal (as opposed to cross-sectional) in nature. Models built to create price or rent indices are a primary example of longitudinal hedonic price models.

Repeat sales models are a special case of a hedonic price model often used to create real estate price indices. As the name suggests, repeat sales models only use data for properties that have sold more than once. This is particularly advantageous when examining price movements in markets with a large amount of housing stock that transacts regularly. Repeat sales are widely used to develop indices that reflect price changes as they compare differences in value for the same property over time. The most well-known repeat sales models are the CoreLogic/S&P Case-Shiller Home Price Indices, which include a 20-city composite index, individual indices for the 20 metropolitan areas, and a 10-city composite index. Economists Karl Case and Robert Shiller developed the first of the indices, tracing U.S. house prices back to 1890, the date to which all the values in the subsequent indices are normalized. Another example of a repeat sales index is the House Price Index methodology proposed by economists at the Federal Housing Finance Agency (Bogin, Doerner, & Larson, 2016a, 2016b, 2016c) in the United States.

Repeat sales are of considerable value when engaging in predictive analytics, but the method is of little use in markets where multiple transactions have not been recorded. Lack of repeated sales can result in reduced or insufficient sample size, potentially limiting the reliability of results.

Diagnostics

One of the primary concerns in any type of analysis is the validity of the data that are used. Data validity refers to the degree to which the data accurately reflect the question the researcher seeks to address; it is important at all levels of the process, affecting not only data collection but also the particular model specifications including the formula and variable transformations. Broadly speaking, diagnostics are a suite of statistical tests that are run on a set of data to detect any threats to validity that can be subsequently rectified or, at the very least, acknowledged and justified.

Multicollinearity

Multicollinearity refers to a condition when two or more variables are highly correlated with each other. This happens quite frequently in research involving the observational data most often used in real estate research. Demographic variables such as income and educational attainment are often highly correlated as are the number of rooms and size of a structure. Predictors created from the data itself (such as second-order polynomial distance variables, or "distance squared") are often strongly correlated with the original measurement. Granted, the latter type of correlation is easily corrected by eliminating one of the measures from the model, but the former is a little trickier to deal with.

The Variance Inflation Factor (VIF) (or its inverse TOL, the "tolerance" metric) is what most researchers use to identify issues with multicollinearity. The VIF estimates the degree that coefficients generated from a regression are inflated due to correlation among variables, but it is important to note that this dependency is not limited to pair-wise comparisons. Multiple variables may exhibit varying degrees of latent correlation that are not readily apparent, and the VIF is one measure that can help sort out these issues.

Statistical packages with user-friendly interfaces such as SAS, SPSS, and Stata include a point and click option for calculating the VIF and some report them by default. Less statistics-friendly software such as Excel require the VIF equation to be manually entered. The VIF is shown as a numeric score with one indicating no correlation and increasing values indicating increasing levels of dependency. Generally speaking, variables with a score greater than four deserve further investigation and those with a score of 10 or more are considered highly correlated.

The question for the researcher, then, is what should be done with variables that are moderately to highly correlated? This is, of course, fully dependent on the aims and objectives of the research, but there are a few general options that the researcher can consider when dealing with the data. The first (and easiest) option is to remove the violating predictor (variable) from the model. Correlation among variables can result in a wide variety of coefficient values for a single variable as different model iterations are produced, thus when the problem variable is removed, coefficients should stabilize, ideally without a substantial impact on the explanatory power of the model. But what if the predictor is one (or more) of those variables that are critical to explaining the research? In that case the researcher can run separate models, excluding the offending variable(s) on a case-by-case basis. Techniques such as stepwise regression and Bayesian Model Averaging (Raftery, Gneiting, Balabdaoui, & Polakowski, 2005) can also assist in the variable selection process.

Both of these options are saddled with issues that make them less than acceptable in many instances. After all, a firm theoretical grounding may dictate the use of certain variables whether they are correlated or not.

Heteroskedasticity

Heteroskedasticity exists when the variance of the relationship between two variables is not constant over the range of the variable(s). The major downside to heteroskedasticity in pricing models is that it creates unreliable impacts of the coefficients of standard errors, potentially leading to erroneous conclusions regarding the statistical significance of a variable(s).

White's test can detect the presence of heteroskedasticity in the error terms of a hedonic (regression) model and offers a method for correcting the standard errors to be more reliable in the face of non-constant variance. Another option is to work with a different functional form of the variable(s). As discussed above, transformations of dependent and/or independent variable may help improve the model specification. Independent variables can also be transformed by creating interaction variables, which are the combination of two or more independent variables to create a new variable. These latent variables are not the same as observed variables (hence the term "latent") but are inferred based on the values of the observed variables. A quality of life index variable may include some combination of crime statistics, income, and school quality data, for example, and may be used to help overcome omitted variable bias. It should be noted that creating interaction variables has the potential to bias the parameter estimates as well as inflate certain VIF statistics, making it important to thoroughly describe and justify any proposed combinations.

Spatial tests and models

The impact of location on value is one of the most important and most difficult tasks in properly specifying a hedonic price model. Within a hedonic modeling framework spatial and location impacts manifest themselves in two primary ways: through (1) spatial heterogeneity; and (2) spatial dependence (Krause & Bitter, 2012). Spatial heterogeneity refers to a situation in which the relationship between the dependent and independent variables vary over space. Spatial dependence, on the other hand, exists when observed values of nearby observations are impacted by each other or by a common latent or omitted variable. More generally, issues of spatial heterogeneity can be seen as "macro-spatial" issues dealing with how best to segment a given geographic area into functioning submarkets, whereby issues of spatial dependence exist as "micro-spatial" issues referring to how to account for all of the highly local influences on prices or rents.

Spatial heterogeneity

One of the more abstract questions associated with real estate is how the boundaries for markets and submarkets are defined. In some research market boundaries are easily defined, especially when they are predetermined by government (e.g., the city, the postcode, the suburb, the school district). Other professionals ascribe to the "you know it when you see it" approach to defining market or submarket boundaries (e.g., the subdivision, homes within 500 meters of a train station, or the change in housing stock from one area to another). While these definitions are often reasonable and defensible, they do not fully capture the dynamics of the relationships within those prescribed boundaries.

The literature provides no clear consensus on how to delineate submarkets. Some researchers argue that submarket boundaries are more accurately defined as a function of structural similarities (Ugarte, Goicoa, & Militino, 2004), characteristics of the neighborhood (Tiebout, 1956; Sieg, Smith, Banzhaf, & Walsh, 2002), stratification of geographic areas (Abraham, Goetzmann, & Wachter, 1994), price (Goodman & Thibodeau, 2007), or demographic and attitudinal similarities among residents who occupy the properties being analyzed (Lipscomb & Farmer, 2005; Farmer & Lipscomb, 2010; Belasco, Farmer, & Lipscomb, 2012). However, data availability is a key driver that leads the researcher to consider methods that require more nuanced or high-resolution data (e.g., demographic and attitudinal data of local residents) or less nuanced data (e.g., sale prices and property characteristics).

From a testing standpoint, a spatial Chow test (Anselin, 1990) may be employed to determine if breaking the data into two or more sub-markets presents an improvement in performance and/or significant differences in the coefficient estimates. Evidence of significant differences in the coefficients between the proposed submarkets may suggest that different market dynamics affect each these various areas and that they should be examined using separate models.

Another approach is to ignore the issue of creating definitive boundaries between submarkets and to estimate more flexible models that allow the relationship between dependent and independent variables to vary smoothly over space. Referred to as Geographically Weighted Regression (GWR) models (Fotheringham, Brunsdon, & Charlton, 2002), these models estimate a separate model at each point in space, with observations weighted based on their distance to the estimate point. The result of a GWR modeling approach is a matrix of coefficient estimates, one for each variable at each estimate point in space. The coefficient can then be mapped to show the "drift" or change in coefficients over space. From here, more traditional submarketing techniques can also be used to divide up the area based on the drift.

Spatial dependence

Once appropriate submarkets are determined, finer scale or micro-spatial issues can be addressed. Tests for spatial autocorrelation (as discussed in Chapters 8 and 9) are the most common approach to determine if unaddressed issues related to locational influences on prices or rents remain unaccounted for in the model. Most often, omitted variables – the failure to include all spatial influencers of price or rent – are a primary cause of high levels of spatial autocorrelation in the error terms of a hedonic price model. If tests for spatial autocorrelation suggest that spatial dependence remains in the model, then one of the following approaches should be pursued to remedy the situation.

Micro-level locational factors (the likely cause of spatial dependence) are notoriously hard to properly specify and control for in a standard hedonic modeling framework. The simplest approach to attempting to control for highly local influences on prices or rents is to use a set of spatial fixed effects or binary indicator variables that reference some small spatial disaggregation such as postal codes or official statistical boundaries (e.g., census tracts). While the ease of estimation for this approach is appealing, its efficacy is directly related to the ability of the pre-defined spatial units to properly capture differences in local amenities. As Koschinsky, Lozano-Gracia, and Piras (2012) show, the use of spatial fixed effects is better than a model without any spatial specification but is inferior to more complex approaches to controlling for spatial dependence.

A more advanced way of controlling for spatial dependence in hedonic price models is through the use of spatial econometric specifications. Spatial econometric models explicitly and simultaneously address the issue of spatial dependence. The most common approaches are spatial lag and spatial error models (Anselin, 1988; Osland, 2010). The spatial lag model assumes that spatial dependence is manifest in the dependent variables – the observed prices or rents – and tackles the issue by estimating the "lag" or direct relationship between nearby observations. Spatial error models, on the other hand, assume that the resulting spatial dependence is manifest in the error term and therefore results from omitted spatial variables. Spatial error models correct for spatial dependence through corrections of the error terms in the model.[1]

To determine which is the more likely form of spatial nuisance, lag or error, a LaGrangian multiplier (LM) test can be used. The results of the LM test will provide an estimate of which correction, error, or lag, is the most appropriate given the data and non-spatial model that had been previously estimated.

Regardless of whether using a lag, error, or more complex spatial econometric specification, the structure of the spatial relationship must be defined. The structure of spatial relationship relates to the extent and strength of spatial relationships between observations. For example, is observation X influenced only by the nearest observation or by the nearest 10 or even 20 observations? Does the strength of the relationship decrease with distance, and if so, by how much? All of these questions are answered by the specification of a spatial weights matrix (SWM), which defines those observations thought to be related to each other as well as the strength of that relationship. There is a certain measure of subjectivity in selecting a spatial weights matrix, and the analyst or researcher should attempt to validate their choice through knowledge of the area or through sensitivity tests.

As the size and geographic scope of datasets have expanded, the importance of controlling for spatial effects – heterogeneity and dependence – has only increased. Unfortunately many of the spatial econometric techniques are computationally heavy and not available in all software packages. Additional computing power and/or different software is often required. As a result, an important component of the analytical workflow and research plan is to develop a strategy for dealing with the possibility for some form of spatial effects in the analytical modeling phase.

Sensitivity tests

Rarely will a complex real estate analysis like a pricing model offer perfect clarity in the decisions regarding discordant value treatment (Chapter 9), model specification including variable transformation, potential subsetting due to submarkets and treatments of spatial autocorrelation, and other issues such as multicollinearity and heteroskedasticity. As a result, it is often prudent to undertake sensitivity tests. A sensitivity test is the re-running of a model with a slight change in parameters to determine if the choice of parameter has a significant impact on the final results.

There are generally four decisions or parameter types over which sensitivity tests are often used: (1) model specification; (2) treatment of discordant values; (3) data subsetting; and (4) data sampling. Sensitivity tests of model specification estimate alternative models, both in terms of the variables included (or excluded) from the model as well the transformations to the variables. For example, if an analyst is uncertain as to whether to impose a log transformation on the dependent variable, they can simply run the model once with it and once without and compare the results. If the resulting estimates are rather similar, then the choice of transformation is somewhat irrelevant or the model is not sensitive to the choice of transformation. If the model is deemed to be sensitive, then more justification and/or research should be conducted to validate the choice of transformation.

Sensitivity tests can also be employed to determine the impact of including or excluding discordant values (outliers or suspected errors) from the analysis. Like the process for model specification, the models are run once with all discordant values in the data set and once without. Differences in model performance and coefficient estimates are then examined between the two (or more) alternative models. If the inclusion of certain discordant values has little to no impact on the final results, it may be prudent to leave those observations in the data to reduce potential claims of researcher bias.

The use of submarkets and the boundaries thereof can be tested for sensitivity as well. The Spatial Chow test discussed earlier is one form of this. Additional tests can be done between different submarket boundaries to determine if the delineation of submarkets has a noticeable impact on market performance and coefficient estimates.

Finally, the impact of data samples on results, usually in regards to predictive accuracy, can be tested for sensitivity with a cross-validation test. Cross-validation is a holdout sensitivity testing procedure that begins with separating a dataset into k partitions, or folds, where k is set to 5 or 10, for example. The cross-validation process begins by iterating k times, holding out one of the k folds as a test set while the rest are combined to inform the model. With each iteration the cross-validation test produces an estimate that can be averaged to generate the overall estimate. Cross-validation tests can be used to judge how well a particular model result holds over a different (smaller) sample of data as well as providing more accurate out-of-sample prediction errors.

Other models

Hedonic price models are not the only approach to modeling real estate prices and rents. Many additional types of models have been suggested and tested over the past few decades, including, but not limited to, artificial neural networks, random forest models, automated comparable sales techniques, and broader machine learning approaches.

Broadly speaking, the Classification and Regression Tree (CART) Methodology, also known as decision trees, can be used to describe a set of complex data and inform decision making. A classification tree creates models where the target variables can only be a set of finite values. The target variables of regression trees, on the other hand, take on continuous values. Within the CART framework, several processes exist to remove unnecessary variables from the process, filter out influential data observations, and reconcile multiple outputs before estimating models.

Artificial Neural Networks (ANN) are sets of classifiers that encompass several layers. The output of the first layer is incorporated into the second layer, and so on and so forth. Between the first and second layer lies hidden layers that are increasingly more complex with the number of hidden layers present. When neural networks have several hidden layers, "deep learning" results and performance improves (Ghani & Schierholz, 2017). While neural networks are becoming increasingly popular, they require a large amount of data and computing power that is not always available.

One of the most influential supervised learning algorithms is the random forest regression. Breiman (1999, 2001) designed the random forest algorithm by borrowing elements from the development of aggregate statistical models through its life-cycle. The notion of the classifier tree itself was based on decision tree methods used in handling survey data by Morgan and Sonquist (1963). Breiman, Friedman, Stone, and Olshen (1984) and Breiman (1996) theorized trees could be used for regression and classification by bagging a randomized subset of data. Random forest models run a large number of decision tree models that classify and categorize observations into small, like groups. The resulting predicted rent or price for each observation is averaged over the number of individual trees that are estimated, and this average or ensemble value is taken as the predicted price or rent. Random forest approaches are most commonly used for prediction-focused modeling projects.

Automated comparable sales models seek to mimic the approach of traditional appraisers (valuers) by using a small set of comparable sales from which to derive price or rent estimates and, potentially, the marginal contribution to price or rent of various property characteristics (adjustments).

Machine learning, a form of artificial intelligence, was first created in 1959 when Arthur Samuel programmed a computer to play checkers. The computer improved its performance with each additional game until it performed better than its human opponents (Ghani & Schierholz, 2017). Machine learning is now infiltrating countless areas of everyday life with cars that can drive and park themselves serving as one well-known example. Machine learning is also used to predict consumer preferences such as product or service recommendations through personalized ads.

Ghani and Schierholz (2017) describe machine learning systems as generally producing a large number of features, or variables, instead of using a small amount of predictors. The machine learning system empirically holds out data to perform model selection where it is customary to have models that incorporate thousands of variables, based on the assumption that it is best to have more simple models that are derived from many features instead of complex models that are derived from a small group of features. Model simplification means that the machine trains, or learns, faster. Machine learning is a growing area that enhances research techniques in several ways including prediction, mistake recognition, text analysis, adaptive surveys, and variable selection.

From a real estate perspective, machine learning can be used to automatically identify comparable sales used for valuation practices including appraisals, AVMs, broker price opinions, and more. Several large real estate research firms and government agencies have used machine learning in its algorithms to predict sale prices.

Role of big data in pricing models

The economic theory underlying hedonic pricing models suggests that consumers are able to differentiate the contribution of various property characteristics and then adjust their bid or willingness to pay for a property accordingly. Using sales or rental transactions to determine the marginal willingness to pay is just a proxy for the actual marginal demand for various characteristics. With the explosion of big data and the computing power to match, analysts may be able to more directly (and accurately) measure consumer demand for certain property as well as specific property characteristics. Additionally, this new found data may also help identify changes in macro-market trends earlier than previously possible.

Search data

Search data comes in various formats and can be used extensively in predictive models. The primary source of publicly available (though limited in scope) search data is Google. Many of the real estate listing portals such as Zillow, Redfin, Domain.com.au, etc., have more targeted housing market search information generated by the users of their sites, but to date this information is held proprietarily and is not widely available.

The tools available through Google are always improving and increasing, as is the body of research using the data that now addresses everything from flu trends to political persuasion to unemployment. Google Analytics is based on the use of touch points, or clicks, to understand consumer behaviors and desires, and the data that are gathered by Google can be processed and shared for additional research. The real estate literature includes at least two examples of studies that use Google Insights for Search to predict the house price-volume relationship in the U.S. and U.K. (Hohenstatt, Kasbauer, & Schafers, 2011; Hohenstatt & Kasbauer, 2014, respectively).

Social media (e.g., Twitter) data

Social media data can be gathered for use in predictive models. Some social media data are available online for no charge with an API while others must be purchased (Neylon, 2017). Twitter and Facebook data are available for free (to an extent) and are accompanied by an API. Tweets can be gathered in a number of ways, but the most effective for real estate research are those that are geocoded or those that refer to or hashtag a specific property-related topic (e.g., interest rates or #houseprice). Facebook data includes the number of likes, posts, and shares associated with a particular web page, while LinkedIn CV-based data can be retrieved along with publications and other project information. These data are sometimes available at no charge while at other times they are fee-based. Altmetric.com, a comprehensive source that encompasses numerous social media and mainstream media outlets, provides aggregated social media data for purchase.

Not surprisingly, the use of social media marketing by real estate agents is already quite high. The recent survey entitled "The Impact of Social Media in Real Estate" suggests that 87 percent of respondents already use social media in their real estate businesses to list properties for sale, to drive web traffic, and for other uses (Better Homes & Gardens Real Estate, 2016). We expect that social media will play an increasingly larger role in real estate as the information becomes more available and easier to work with.

Web scraped data

Web scraped data for predictive models can be acquired by using a program to download and process web pages to gather the underlying information. While web scraping is an effective way of gathering large datasets, there are several limitations to the process. Many websites simply prohibit data scraping while others make it very difficult to gather the data. A more efficient way to gather web data is through organization websites. Many even provide APIs that enable scripted and programmatic access to their data (Neylon, 2017). We discuss the ethics and ownership issues with web scraped data in more detail in Chapter 13.

Computational tools

Google, Amazon, and Microsoft offer several big data products and services. Google's Big Data Solutions provides a serverless platform for its applications, and BigQuery is Google Analytics' warehouse that users can access using SQL in a pay-as-you-go format. Data Studio allows users to store, clean, and customize datasets in one user-friendly location. Google Cloud Dataprep is a service-based tool that is used to clean and prepare data for analysis, and the Cloud Prediction API provides a RESTful API to build machine learning models that create tools to enhance data analysis. Amazon offers competitive products through its consolidated Amazon Web Services (AWS) platform including S3 for on-demand storage, Redshift for data warehousing, Kinesis for streaming data analysis, and EMR for a Hadoop and Spark framework. Similar options can be found under Microsoft's Azure platform, including storage, integration, visualization, and integration. These are only a few of examples of big data analysis tools; however, they are most widely used and likely the most useful for very large-scale predictive models in real estate analyses.

Conclusion

Note that pricing models are just one, though probably the most common, use case of real estate analysis in the information age. What is covered in this chapter is simply a broad overview of the pricing model process; hundreds of published journal articles, books, and manuals exist that dive much deeper into the specifics of model specification, diagnostics, spatial models, and sensitivity testing.

Note

1 The reader is directed to Anselin (1988) or LeSage and Pace (2009) for a complete overview of spatial econometric specifications.

References

Abraham, J. M., Goetzmann, W. N., & Wachter, S. M. (1994). Homogeneous groupings in metropolitan housing markets. *Journal of Housing Economics*, *3*(3), 186–206.

Anselin, L. (1988). *Spatial Econometrics: Methods and Models.* Dordrecht: Springer.

Anselin, L. (1990). Spatial dependence and spatial structural instability in applied regression analysis. *Journal of Regional Science*, *30*, 185–207.

Belasco, E., Farmer, M. C., & Lipscomb, C. A. (2012). Using a finite mixture model of heterogeneous households to delineate housing submarkets. *Journal of Real Estate Research*, *34*(4), 577–594.

Better Homes and Gardens Real Estate. (2016). The Current State of Social Media Marketing in Real Estate. http://bhgrealestateblog.com/current-state-social-media-marketing-in-real-estate/, accessed on 23 May 2017.

Bogin, A. N., Doerner, W. M., & Larson, W. D. (2016a). *Local house price dynamics: New indices and stylized facts.* Federal Housing Finance Agency Working Paper No. 16–01.

Bogin, A. N., Doerner, W. M., & Larson, W. D. (2016b). *Local house price growth accelerations.* Federal Housing Finance Agency Working Paper No. 16–02.

Bogin, A. N., Doerner, W. M., & Larson, W. D. (2016c). *House price index accuracy and mortgage credit modeling.* Federal Housing Finance Agency Working Paper No. 16–04.

Breiman, L. (1996). Bagging predictors. *Machine Learning*, *24*(2), 123–140.

Breiman, L. (1999). *Random forests – random features.* UC Berkeley, Statistics Department, Technical Report 567. Berkeley, CA.

Breiman, L. (2001). Random forests. *Machine Learning*, *45*(1), 5–32.

Breiman, L., Friedman, J., Stone, C. J., & Olshen, R. A. (1984). *Classification and Regression Trees.* Boca Raton, FL; CRC Press.

Des Rosiers, F. & Theriault, M. (2008). Mass appraisal, hedonic price modeling and urban externalities: Understanding property value shaping processes. In T. Kauko, & M. d'Amato (Eds.), *Mass Appraisal Methods: An International Perspective for Property Valuers* (pp. 111–147). West Sussex, UK: Wiley.

Farmer, M. C. & Lipscomb, C. A. (2010). Using quantile regression to reveal hedonic submarket competition. *Journal of Real Estate Research*, *32*(4), 435–460.

Fotheringham, A. S., Brunsdon, C., & Charlton, M. (2002). *Geographically Weighted Regression: The Analysis of Spatially Varying Relationships.* Chichester, UK: Wiley.

Ghani, R. & Schierholz, M. (2017). Machine learning. In Foster et al. (Eds.), *Big Data and Social Science* (pp. 147–186). Boca Raton, FL: CRC Press.

Goodman, A. C. & Thibodeau, T. G. (2007). The spatial proximity of metropolitan area housing submarkets. *Real Estate Economics*, *35*(2), 209–232.

Hohenstatt, R. & Kasbauer, M. (2014). Geco's weather forecast for the U.K. housing market: To what extent can we rely on Google Econometrics? *Journal of Real Estate Research*, *36*(2), 253–281.

Hohenstatt, R., Kasbauer, M., & Schafers, W. (2011). "Geco" and its potential for real estate research: Evidence from the U.S. housing market. *Journal of Real Estate Research*, *33*(4), 471–506.

Koschinsky, J., Lozano-Gracia, N., & Piras, G. (2012). The welfare benefit of a home's location: An empirical comparison of spatial and non-spatial model estimates. *Journal of Geographical Systems*, *14*(3), 319–356. http://doi.org/10.1007/s10109-011-0148-6.

Krause, A. & Bitter, C. (2012). Spatial econometrics, land values and sustainability: Trends in real estate valuation research. *Cities*, *29*(S1), S19–S25.

Lancaster, K. J. (1966). A new approach to consumer theory. *Journal of Political Economy*, *74*(2), 132–157.

Lesage, J. P. & Pace, R. K. (2009). *Introduction to Spatial Econometrics.* Boca Raton, FL: CRC Press.

Lipscomb, C. A. & Farmer, M. C. (2005). Household diversity and market segmentation within a single neighborhood. *Annals of Regional Science*, *39*, 791–810.

McMillen, D. P. & Dombrow, J. (2001). A flexible Fourier approach to repeat sales price indexes. *Real Estate Economics*, *29*(2), 207–225.

Morgan, J. N. & Sonquist, J. A. (1963). Problems in the analysis of survey data, and a proposal. *Journal of the American Statistical Association*, *58*(302), 415–434.

Neylon, C. (2017). Working with web data and APIs. In Foster, I., Ghani, R., Jarmin, R., Kreuter, F. and Lane, J. (Eds.), *Big Data and Social Science* (pp. 23–70). Boca Raton, FL: CRC Press.

Osland, L. (2010). An application of spatial econometrics in relation to hedonic house price modeling. *Journal of Real Estate Research*, *32*(3), 289–320.

Raftery, A., Gneiting, T., Balabdaoui, F., & Polakowski, M. (2005). Using Bayesian model averaging to calibrate forecast ensembles. *Monthly Weather Review*, *133*, 1155–1174.

Rosen, S. (1974). Hedonic prices and implicit markets: Product differentiation in pure competition. *Journal of Political Economy*, *82*(1), 34–55.

Sieg, H., Smith, V. K., Banzhaf, H. S., & Walsh, R. (2002). Interjurisdictional housing prices in locational equilibrium. *Journal of Urban Economics*, *52*, 131–153.

Tiebout, C. (1956). A pure theory of local expenditures. *Journal of Political Economy*, *64*(5), 416–424.

Ugarte, M. D., Goicoa, T., & Militino, A. F. (2004). Searching for housing submarkets using mixtures of linear models. In J. P. LeSage, & R. K. Pace (Eds.), *Spatial and Spatiotemporal Econometrics* (pp. 259–276). Oxford: Elsevier.

11 Analysis of unstructured text

Many of the studies published in real estate journals have historically focused on the investment potential of property. Research on REITs, property valuation, pricing models, and mortgage choice are quite overt in this assumption, and even much of the research that focuses on real estate as a consumption good (i.e. housing preference) holds an inherent bias toward investment, though less explicitly so. Because of this, the vast majority of real estate studies are quantitative in nature and related to the financial impacts affecting stakeholders such as investors, homeowners, neighborhoods, and/or cities.

Studies using qualitative data generally do not deviate far from this, though the use of qualitative data in real estate is far less common and tends to be arguably less revered in academic circles. Over the past several years, however, there has been a shift in real estate research as it has drifted toward a more behavioral perspective following similar trends in economics and finance. This has led to an increased interest in qualitative data as well as the methodologies and software that support them. While the emphasis on real estate as an investment remains, the tools have evolved to support a more holistic view of the discipline.

This chapter focuses on some of these new phenomena that saw an explosion of activity after 2010 with the emergence of digital humanities (Ignatow & Mihalcea, 2017, p. 160), specifically covering various forms of content analysis related to the analysis of unstructured text, also known as *text mining* or *text analytics*. We begin by exploring the methods that have been recently tested in the real estate literature and highlight other promising ways of examining the behavioral characteristics of real estate such as sentiment analysis. We detail the importance of semantic analysis and conclude the chapter by introducing Latent Semantic Analysis, an emerging method used to evaluate large bodies of text.

Content analysis in real estate literature

Broadly speaking, content analysis is about increasing knowledge through the systematic review and evaluation of textual data. As an analytical technique, it can be used to evaluate a wide variety of data including written documents, oral communications, and graphics with the aim of making replicable and valid inferences through the interpretation and coding of text. Computers have streamlined and simplified the process, which culminates in the conversion of qualitative data into quantitative data that can be combined with other forms of data to substantively enhance knowledge on a given subject.

There are few disciplines that have not benefited from work in this area. Studies of firm behavior, media coverage, social movements, decision-making, and patient care are rife throughout the literature as researchers look to harness the large amounts of textual data that are now digitally available. These data are much more accessible and easier to navigate than qualitative data have ever been, and because dealing with large volumes of data is progressively becoming less of an issue, the demand for it continues to grow.

Inductive content analysis

Early attempts at content analysis in real estate were conducted without computers and primarily involved the use of Inductive Content Analysis (ICA) focused on literature reviews and, to a lesser extent, responses from open-ended questionnaires. Inductive content analysis relies on a researcher's inductive reasoning skills to sort through data, review and classify it according to themes. Results are derived from open coding wherein data are repeatedly reviewed, reassessed, and regrouped. Progressively larger groups representing broader themes are created at each step.

Examples of the use of ICA in the literature include a study of the value-adding components of real estate to corporate wealth based on information extracted from 39 responses to an open-ended survey – a small study from which eight main themes were derived (Lindholm, Gibler, & Levainen, 2006). The majority of other examples are rooted in literature reviews where papers are examined and placed into categories. Johnson, Roulac, and Followill (1996) examined the content of papers presented during the first 10 years of the American Real Estate Society annual meetings, identifying the 16 most frequently researched and presented topics including urban economics/development, valuation, mortgage applications, international, corporate real estate/asset management, brokerage, investments, risk/portfolio management, taxes, securitization, body of knowledge, cases, legal issues, computer applications, markets and institutions, and government programs. Jud (1996) reviewed nine years of published research in the *Journal of Real Estate Research* and noted 10 primary subject areas including investment, appraisal, corporate property, housing values, brokerage, REITs, housing markets, regional, environment, and mortgages. Dombrow and Turnbull (2004) extracted nine topics – appraisal, brokerage, housing, institutions, investments, mortgages, nonresidential, public policy, and other – after reviewing all the papers published in the *Journal of Real Estate Finance and Economics* and *Real Estate Economics* from 1988 to 2001.

One of the more interesting components of these studies, when observed as a whole, is just how little consistency there is between the topics identified by the researchers. Granted, each of the studies used different sets of data and there are some consistencies, but one would expect to find a fairly standard set of topics that represent a specific discipline. This highlights some of the difficulties in using methods that are more subjective than objective and is particularly noticeable in the two studies published in 1996 that cover nearly the same timeframe.

Sentiment analysis

Understanding how investors feel about a certain asset class is fundamental to understanding how the market for that asset will change over time. Increases (decreases) in stock prices, for example, are predicated on how investors perceive the information that is available for that specific investment. Sentiment analysis investigates these perceptions, feelings, and beliefs to assess the prevailing mood of investors, making qualitative judgments to predict directionality. Sentiment studies are focused on the general tone that impacts the market as opposed to the effects of a single incident or release of information (event studies).

Market sentiment studies are fairly pervasive throughout the literature (Clayton, Ling, & Naranjo, 2009 among others), but what makes research in the big data era different is that quarterly reports, surveys, and other materials with generous time lags are no longer necessary to measure market sentiment. That information is now available in real-time, is continuously being produced, and participation is relatively unconstrained and non-selective. News analytics (focused on news cycles), semantic analysis, and other linguistic techniques are used in the evaluation.

Arguably the most well-known market sentiment study of the big data era looks at the use of Twitter feeds to predict movements in the Dow Jones Industrial Average (DJIA). Bollen, Mao, and Zeng (2011) incorporated two mood-tracking tools, OpinionFinder and Google-Profile of Mood States, to investigate the effect of two general sentiments (positive and negative) and six different moods including calm, alert, sure, vital, kind and happy, on the closing price of the DJIA. Using a Granger causality analysis and a Self-Organising Fuzzy Neural Network, the study found that the direction of the change in the closing value could be predicted, with 87.6% accuracy, within three to four days, specifically related to the mood "calm." None of the other moods or the broad sentiment categories produced significant results.

The REIT marketplace provides an open portal for research into sentiment analysis in much the same way as the stock markets do. REITs are highly liquid, trade continuously, and are very sensitive to information even though the underlying assets are fixed. Existing research on sentiment analysis in the REIT market typically employs one of two methods to understand the impact of investor sentiment on prices. Older studies tend to resist making predictive judgments about directionality, instead relying on price movements to indicate whether sentiment is positive (prices increase) or negative (prices decrease). Downs, Guner, Hartzell, and Torres (2001) studied the impact of market commentary on REIT prices by looking at unexpected price and volume movements that occurred on days when a well-respected Barron's column, The Ground Floor, released REIT-specific information. The objective of the paper was to determine if the commentary played a significant enough role in market perceptions that it influenced investment behavior. The interesting component relative to sentiment analysis is that the authors specifically sought to remove qualitative assumptions about the directionality of the effect, assuming that any movements would result in a rebalancing of portfolios.

The Downs et al. research contrasts with recent studies that analyze sentiment in an effort to predict the direction of market movements. Doran, Peterson, and Price (2012) looked at a similar issue as Downs et al., albeit in a predictive manner, by analyzing the impact of conference call dialogue on REIT prices. The calls were examined in their entirety with particular emphasis put on the question and answer portion under the assumption that this is where information about transactions outside of quarterly earnings is often expressed. The tone or sentiment of the discussions between management and analysts were also probed and comparative measures of positive, negative, active, passive, strong, weak, overstated, and understated applied. The results of the study show a significant, positive relationship between tone and abnormal returns, i.e., the more positive the tone the higher the return. Results further show that a strong positive tone almost completely offsets a negative earnings surprise, making a convincing argument that the way information is delivered has the potential to override the fundamentals of a security when it comes to investment decision-making, at least in the short term.

Sentiment studies like the Bollen et al. work assess relatively short-term moves in asset prices thus comparable examples applied to fixed assets like real estate are quite rare. Real estate prices are "sticky," i.e., market values typically do not fluctuate in time to the same degree as other, more liquid assets. Income streams are subject to contractual obligations, locking in rents for extended time periods, and expecting residential prices to react in the short time frame of the Bollen et al. study is problematic, given the 30-plus day period between contract and settlement dates of most transactions. Moods extracted from social media can change relatively frequently given any number of conditions, making connections to fixed property prices difficult to establish.

Sticky prices mean that fixed assets need to be approached from a different vantage rather than conform to the framework established for sentiment studies of equities and REITs. Relating sentiment to the geographic distribution of real estate prices is one approach, as prices vary substantially from market to market. The Trulia web site[1] (United States) includes a heat map[2] that displays median home sales prices per square foot at the county level, with a range from $26 (for Bent County in Colorado) to $950 (for San Francisco County in California). The price range increases at the zip code level, with the most expensive locations in New York City/Lower Manhattan in New York, Aspen in Colorado, Los Angeles/Beverly Hills in California, and San Francisco/The Presidio, in California, where median listed home prices exceed $4 million. In chapter 12, we provide a case study that examines the relationship between the geographic distribution of home prices and local sentiment in the city of Seattle, Washington.

Semantic analysis

Semantic analysis is rooted in the multiple notions that (1) human communities, cultures, organizations, disciplines, etc., organize their thoughts and communications to create meaning; (2) that meaning is subject to contextual variations; and (3) words are used in combination to develop language that defines those communities and becomes part of their identity. The first assumption relates to theories of order creation. The second is best exemplified in the property discipline by simply considering the term "real." Real estate, real income, real property, real rate of interest – these are just some of the combinations of terms in the discipline that involve the term "real." When combined with "estate" and "property," "real" defines a fixed asset including all the implicit rights and obligations that go with the asset. Real incomes and real interest rates are incomes and rates that take into account the effects of inflation. All four concepts are important to real estate research, yet the term "real" implies something quite different in combination with other words.

The third assumption takes into account the overall structure as well as the individual components of the language associated with a particular community, culture, or discipline. Because real estate is multi-disciplinary by nature, it makes sense to define it using terminology sourced from economics and finance as well sociology and ecology. Words and concepts like inflation, mortgage, demographics, and brownfield have their foundations in those disciplines, respectively, yet they also play an integral role in the structure of the real estate "language."

To put it succinctly, semantic analysis is the attempt to derive meaning from the study of a collection of related texts or statements that include words, combination of words and phrases. Texts are related in that, even though they are individually prepared or created based on the thoughts of an individual or group of individuals, there is a common cognitive thread tying them all together. Examples include studies of REIT annual reports where each REIT is legally obligated to prepare a report describing yearly performance, surveys with open-ended questions completed by respondents who possess knowledge of the issues under question, and consumer reviews of a single product written by multiple, otherwise unrelated, people, who share an experience through that product. We will revisit this point later in this chapter, when we examine the content of property descriptions written by people, and for people, who share the common interest in browsing through properties listed for sale in the same local market.

Two of the most commonly used algorithmic approaches to semantic analysis (see, e.g., Ignatow & Mihalcea, 2017, p. 158; Larsen & Bong, 2016) are Latent Semantic Analysis (LSA) and Latent Dirichlet Allocation (LDA). From the statistical estimation viewpoint, LSA uses a dimensionality reduction approach similar to principal component analysis, and LDA uses a Bayesian estimation method based on Markov Chain Monte Carlo simulations. From a cognitive science viewpoint, LSA simulates mental processes followed by a human reader who would read a collection of text documents, notice the patterns of word co-occurrence, and attempt to organize document meaning using hierarchies of topics. In contrast, LDA simulates the mental process followed by a human author who, already being aware of the topics, would compose each document by selecting associated topics within the documents and associated words within the topics. We now focus our discussion on LSA, which is computationally more efficient, generates stable results that can be replicated, and is currently at the forefront of those methods used in the real estate discipline.

Latent semantic analysis

LSA (also known as Latent Semantic Indexing) uses a combination of data preparation and semantic extraction techniques to identify the topical structure contained in large bodies of text by identifying patterns in the language that is used. LSA extracts socially constructed components of meaning by identifying two types of variables, the terms (words) and documents in which those words are contained (Landauer, 2007). Because it is systematic, iterative and able to process large bodies of work via computer applications, LSA is an excellent vehicle for evaluating the intellectual content of textual data. It is a multilayered process that can be used to categorize structured textual data or extract latent themes from unstructured forms of data with a textual component.

There are several examples of LSA used in the real estate literature, primarily stemming from research conducted by Winson-Geideman and Evangelopoulos in their study of the topical structure of the real estate discipline from 1973 through 2010 (Winson-Geideman & Evangelopoulos, 2013a, 2013b, 2013c). All three studies are derived from a common set of data that includes all of the abstracts published in the top real estate journals – *Journal of Real Estate Research, Real Estate Economics*, and *Journal of Real Estate Finance and Economics* – from 1973 through 2010. By focusing on the content of the abstracts the authors identified subject categories and placed the respective articles into the category that best represented the content of the abstract and hence the research.

What set the Winson-Geideman and Evangelopoulos research apart from earlier attempts to categorize real estate research is that (1) the subject categories were developed directly from the research in a systematic, bottom-up fashion; and (2) the placement of the articles into specific subject categories was organically induced as the articles themselves dictated the subject classification. This is in contrast to other studies that used a top-down approach with a predetermined topical structure that research papers were subsequently categorized into (Harrison & Manning, 2008). While we provide a brief summary of the

LSA process in the next section, we encourage the reader who is considering text analysis to also turn their attention to the *Journal of Real Estate Literature* where Evangelopoulos, Ashton, Winson-Geideman, and Roulac (2015) provide a more thorough guide to the application that includes code and software recommendations.

The LSA process – a brief summary

As with any project, the research process begins with the identification of a problem that leads to the development of a research question. Once a research question has been established and the data source(s) identified, the next step is to determine whether or not LSA is the appropriate method for analyzing the data. If the data are textual and if the research question involves understanding the content of the text to derive a meaningful response to the question, then LSA is a likely candidate for the analysis.

Analyzing documents with LSA involves the extraction of latent topics or ideas and correlating those ideas with the specific terms and documents from which they are drawn. The body of literature that is collected and relevant to the research is known as the *corpus* (plural: corpora). The corpus can be quite large, consisting of multi-page documents or quite small, consisting of a briefly worded response to an open-ended question. The unifying, socio-cognitive relationship between the documents within the corpus should be intuitive and not random, e.g., Tweets containing the phrase, "buy a house" or the comment section of an editorial about house prices in a newspaper. The first example includes a clear relationship to an external unifying parameter (the phrase "buy a house") whereas in the second example the comments flow from a directly related item (the editorial about house prices).

The data can be collected and stored in a variety of different formats, depending on how much and what type of data are being collected. The data set for the Winson-Geideman and Evangelopoulos study, for example, included 2,526 abstracts downloaded into Excel with title, author, volume, journal, and abstract vectors. Larger volumes of data can be stored in Microsoft Access or another type of software subject to the requirements of the research.

To properly prepare the data for analysis, we turn to the Vector Space Model (Salton, 1975), which dictates the best procedures for cleaning textual data and reducing them to their most usable and relevant form. The data are subject to a number of different processes including tokenization, filtering, stemming, term weighting, and dimensionality reduction (Coussement & Van den Poel, 2008). *Tokenization* refers to the parsing of the body of text into individual words, using certain characters, such as space, comma, period, etc., as delimiters. *Filtering* refers to the removal of words that contribute little to overall meaning. Words such as "a," "the," "to," and "of" are part of a list of 543 standard "stop words" considered trivial language elements that are generally irrelevant to most topic extraction applications and are therefore excluded from the analysis. A custom dictionary of stop words is also often created, containing those terms that are frequently occurring and unlikely to make any meaningful contribution to the work. For example, the terms "real" and "estate" are left out of real estate studies. The remaining terms are *stemmed*, that is reduced to their root form. The words taxation and taxes, for example, both become *tax–*.

To further reduce term dimensionality, *communality filtering* is applied. This step begins with a count of the number of times a term appears in a document followed by the compilation of a term frequency matrix. Singular Value Decomposition (SVD) is then performed on the matrix, converting it to a term-by-factor matrix that includes term eigenvectors as well as a document matrix that includes document eigenvectors. An initial SVD pre-processing of the stemmed terms identifies those that explain the least amount of variance in the documents, with the remainder representing the components of primary interest to the research. In some cases this step is repeated to further narrow the components with the greatest explanatory value. In general, filtering operations simulate the mental filtering process performed by humans as they try to focus on salient components of meaning by tuning out words and contexts that are not related to a particular cognitive purpose.

After filtering, the remaining terms are weighted based on the number of occurrences within the corpus. *Term weighting* promotes terms that occur less frequently and demotes those that occur more often in an effort to identify the most important and descriptive concepts contained in the corpus. While most LSA-based text analytic software includes some type of weighting method, the one most commonly recommended for business-related data are term-frequency multiplied by inverse document frequency (tf-idf) and the log transformation of entropy (Evangelopoulos, Zhang, & Prybutok, 2012). Term weighting simulates an aspect of human cognition that also emerges in the analysis of stock prices: Fechner's law of psychophysics, introduced in 1860, which states that human perception is proportional to the logarithm of the stimulus intensity. We see applications of this law in the ways humans evaluate information: regardless of the absolute value of the information level, identical percentage increases "feel" the same. Similarly, in the analysis of text, the subjective significance of the presence of a certain word in the document appears to be proportional to the logarithm of its total frequency in the entire document collection.

The final data preparation step involves *dimensionality reduction*. In this step the researcher looks to define the number of dimensions, topics or themes that best address the research question. Some research compares various levels of aggregation (Winson-Geideman & Evangelopoulos, 2013a) where solutions are generated at high levels of semantic aggregation, e.g., a two-factor solution that groups results into two broad topics. Other solutions disaggregate the data into much lower, detailed levels. A more quantitative approach involves examining a scree plot of eigenvalues to identify the point of diminishing return where there is an "elbow" point on the graph, or a change in the slope of the curve.

Regardless of how the researcher chooses to determine the number of dimensions, they must correspond with the aim of the original research question. In the Winson-Geideman and Evangelopoulos (2013a) study the researchers examined topics with levels of aggregation as large as 50 topics but after examining the LSA output settled on 25 as the number that best represented the intent of the study and the results of the analysis. Rotating and labeling the factors to ensure a strong relationship between the number of dimensions and the interpretation of the factors will help validate the choice of dimensionality.

The final step in the process is to compare the topics that have been extracted to the terms and documents from which they are derived. The LSA output shows the terms and documents that load on a specific topic (factor) as well as the factor loadings for each. This creates a hierarchy of terms and documents where the higher the loading, the greater the relevancy of the term and document to that particular factor and the greater the likelihood that that term or document will influence the final factor label. Even though terms and documents with low factor loadings should be less influential in the final label assignments, all the results should be thoroughly examined to make the label as concise and descriptive as possible without containing too much detail.

We now continue with an example, designed to demonstrate the value-adding capabilities of unstructured text. The example focuses on a typical hedonic regression model for home prices. We start by fitting a model that uses traditional predictors, such as the total home area in square feet, the number of bedrooms, the number of bathrooms, and qualitative indicator variables for the neighborhoods included in the data set. We then expand the model to include a non-traditional element, the property description in unstructured text format.

Fitting unstructured text to the hedonic model[3]

As discussed in Chapter 10, a regression analysis is one of the more traditional statistical tools used in real estate. A regression model is typically used to explain or predict a response variable; in the case of residential real estate, the price of a home provides a natural choice for the response variable. As hedonic theory implies, the model uses the implicit value of various home attributes to explain why some homes are more expensive than others and, presumably, more desirable. From the predictive point of view, a regression model predicts a home's price, given the values of the home's individual attributes.

To give an example of how unstructured text can aid in analyzing real estate markets, we provide a fictitious example. We examine a manufactured dataset of 79 home sales from two neighborhoods in the same city, Greenville and Briarwood. Our data contains basic information about the properties as well as narrative description of the property, such as might be entered by the agent or broker. Table 11.1 shows the first three rows in the data set. A description of the variables is shown in Table 11.2.

Table 11.1 Example data set: first three rows

HomeID	*Area*	*Price*	*Sq ft*	*Beds*	*Baths*	*Year*	*Age*	*A*	*Simple Description*
H01	W	700	2560	4	2	1960	57	0	Great Briarwood home with a nearby park
H02	W	777	2120	3	2.5	1990	27	0	Spectacular home with dramatic architecture, and great bayview location
H03	W	447	754	1	1	2002	15	0	Great condo with a nearby park

Table 11.2 Example data set: description of the original variables

Original variable	*Variable description*
HomeID	Home ID, H01-H79
Area	Two neighborhoods: W = Briarwood, A = Greenville
Price	Listed selling price in 2017 (in 1000s of U.S. dollars)
Sqft	Total home area (in square feet)
Beds	Number of bedrooms
Baths	Number of bathrooms
Year	Year when home was build
Age	Age, derived from Year using Age = 2017–Year
A	Dummy variable, recoded from Area. A = 0 if Area = Briarwood, A = 1 if Area = Greenville
Simple Description	Simple text that simulates an MLS property description

Since the purpose of this illustration is to make a case for the analytical value of the property description field, which traditional quantitative analysis would overlook, it is important to fit a good initial model that explains as much variability as possible without using the descriptions. Given the small number of submarkets in our dataset (two), we introduce interaction terms to help control for potential spatial heterogeneity. Our initial model for *Price* will, then, have the following structure:

$$Price = b_0 + b_1 A + b_2(Square\ Footage) + b_3(A \times Square\ Footage) + b_4(Bedrooms) + b_5(A \times Bedrooms) + b_6(Bathrooms) + b_7(A \times Bathrooms) + b_8(Age) + b_9(A \times Age) + \text{Residual}.$$

In this generic model, the coefficients have the following interpretation:

b_0 price of the unbuilt lot in the baseline location of Briarwood (where A = 0);
b_1 flat location premium for homes located in Greenville, where A = 1;
b_2 price of a squared footage unit in the baseline location Briarwood (where A = 0);
b_3 square footage unit price modifier for homes located in Greenville;
b_4 premium for an extra bedroom in Briarwood;
b_5 extra bedroom premium modifier for homes located in Greenville;
b_6 premium for an extra bathroom in Briarwood;
b_7 extra bathroom premium modifier for homes located in Greenville;
b_8 age appreciation/depreciation coefficient (negative if homes get depreciated);
b_9 depreciation coefficient modifier for homes located in Greenville.

The residual value represents the deviation of an individual home from the expected value, due to individual characteristics that make a particular home either more desirable (for positive residuals) or less desirable (for negative residuals) than its observed characteristics would suggest.

As shown in Table 11.3, the regression model explains 82.16% of variability in *Price*, with the normality and constant variance assumptions marginally satisfied. The model includes *Sqft*, *A*Sqft*, and *A*Age* as significant predictors for *Price*. The OLS model coefficient estimates are shown on Table 11.3.

Property descriptions: coding the presence of characteristic terms

After fitting the base model, we now turn to the property descriptions. The critical question is, can we fit a better model and thus explain more variation in the house prices through the addition of qualitative variables extracted from the description? This is accomplished in two stages. First, we introduce indicator variables that account for the

Table 11.3 Regression coefficients for model 11.1

Variable	*Estimate*	*Std. Error*	*t value*	*Pr(>|t|)*
Intercept	211.6	69.1	3.06	0.003
Sqft	0.2253	0.0360	6.25	0
A * Sqft	0.2966	0.0346	8.56	0
A * Age	–4.63	1.02	–4.56	0

Table 11.4 Simple description coding using characteristic terms

ID	*Simple Description*	*Spectacular*	*Location*	*Park*	*bayview*	*Briarwood*	*Condo*
H01	Great Briarwood home with a nearby park	0	0	1	0	1	0
H02	Spectacular home with dramatic architecture, and great bayview location	1	0	0	1	0	0
H03	Great condo with a nearby park	0	0	1	0	0	1

Table 11.5 Regression coefficients for model 11.2

Variable	*Estimate*	*Std. Error*	*t value*	*Pr(>|t|)*
Intercept	160.3	62.9	2.55	0.013
Sqft	0.2107	0.0358	5.89	0
A * Sqft	0.2585	0.0290	8.92	0
A * Age^2	-0.0391	0.0081	–4.8	0
Park * Sqft	0.0485	0.0239	2.03	0.046
bayview * Sqft	0.0777	0.0285	2.72	0.008

presence of certain words or simple phrases. Second, we analyze the full text and perform topic extraction.

Table 11.4 shows the first three rows in the augmented data set, which includes information on the presence of certain terms in the description text. The terms, identified manually, include *spectacular, location, park, bayview* (= a coastal area), *Briarwood*, and *condo* (= condominium, or apartment). With the help of the additional variables, we are able to fit a regression model (Model 11.2) that explains 84.28% of variability in *Price*. The model includes *Sqft*, *A***Sqft*, *A***Age*^{2}, *Park***Sqft*, and *bayview***Sqft*, as predictors for *Price*. This model represents an improvement over the previous model, making the case that there is some useful information in the property descriptions. The OLS model coefficient estimates are shown on Table 11.5. We interpret the model coefficients as follows: the price per square foot increases by $49 if "park" is mentioned in the property description, and/or by $78 if "bayview" is mentioned. In the Greenville area, homes depreciate quadratically over time, while in Briarwood there is no significant depreciation.

Topic extraction using LSA

To extract deeper information from the listing descriptions, we apply latent semantic analysis using the Text Miner™ module in SAS® Enterprise Miner™ (SAS TM).[4] Four topics are extracted, and the results produced by SAS TM include lists of related terms and documents for each topic. Through an examination of those lists, we are able to determine that the generated topics are *nice Greenville condo* (*T*1), *great Briarwood home* (*T*2), *spectacular bayview location* (*T*3), and *nearby park* (*T*4). Topic labels and corresponding number of related homes are listed in Table 11.6.

We then extract the four topics and create four new indicator variables that code each property description by none, one, or more of the extracted four topics. These can be now added as predictors to our regression model.

Table 11.6 Extracted topic labels

Topic	*Label*	*Documents*
T1	Nice Greenville condominium	11
T2	Great Briarwood home	16
T3	Spectacular bayview location	13
T4	Nearby park	19

Table 11.7 Regression coefficients for model 11.3

Variable	*Estimate*	*Std. Error*	*t value*	*Pr(>|t|)*
Intercept	238	64.5	3.69	0
Sqft	0.2138	0.0336	6.36	0
A * Sqft	0.3161	0.0326	9.69	0
A * Age	–4.5950	0.9430	–4.87	0
T1 * Sqft	–0.1521	0.0422	–3.6	0.001

With the help of the additional variables, we are able to fit a regression model (Model 11.3) that explains 84.82% of variability in *Price*. The model includes *Sqft*, *A***Sqft*, *A***Age*, and *T1***Sqft*, as predictors for *Price*. This model represents an additional improvement over the previous model, making the case that the information in the property descriptions is coded better using LSA topics, than using the presence of simple terms. In addition, LSA topics were identified automatically, while simple terms were introduced manually.

The OLS model coefficient estimates are shown on Table 11.7. We interpret the model coefficients as follows: these prices per square foot decrease by $152 if *T1* (= 'nice Greenville condominium') is mentioned in the property description. Greenville homes depreciate by about $4,595 per year, while in Briarwood there is no significant depreciation. We notice that models 11.1 and 11.3 are very similar, but the addition of the interaction term *T1***Sqft* improves the explained variance by 2.66 percentage units because now the model accounts for Greenville homes that are condos.

A final word

To conclude this section we challenge the reader to think about the "old" questions in real estate and to explore new ways to analyze them. While data availability should not be the driving force in research, the proliferation and availability of new types and sources of unstructured and textual data coupled with new modes of analysis make for some exciting research propositions. Some of the new sources (e.g., Twitter) have almost nothing to do with real estate but what they provide are organic, spontaneous, and to a degree unrestricted insights into human thoughts, attitudes, and behaviors. The era of big data offers the researcher the unprecedented opportunity to perform field studies of human sentiment, issues of interest, and their prioritization. Research that links this information to useful outcomes for investors, buyers and sellers of property has the potential to provide enormous gains for academia and industry.

Copyright notice

Notes

1 www.trulia.com/home_prices/, accessed on 25 May 2017.
2 Map data © 2017 Google, INEGI.
3 The data, code, and instructions to replicate this example are available in the online supplement for this book at www.github.com/REAIABook/REAIABook.
4 For those readers who are familiar with numerical factor analysis, topic extraction as implemented by SAS TM combines singular value decomposition with factor analysis rotations (in SAS language terms, *Proc SVD* and *Proc FACTOR*), to produce interpretable topics. It is, therefore, an implementation of LSA. Another way to implement LSA would be by working in the R environment, combining functionality from libraries *tm*, *slam*, *lsa*, and *psych*. In R, the strategy would be to use library *tm* to convert the original text data into a term by document frequency matrix, use library *lsa* to perform the singular value decomposition, and use function *varimax* to rotate the term loadings. For more details, see Wild (2015) and Appendix B in Evangelopoulos et al. (2015).

References

Bollen, J., Mao, H., & Zeng, X. J. (2011). Twitter mood predicts the stock market. *Journal of Computational Science*, *2*(1), 1–8.

Clayton, J., Ling, D. C., & Naranjo, A. (2009). Commercial real estate valuation: Fundamentals versus investor sentiment. *Journal of Real Estate Finance and Economics*, *38*(1), 5–37.

Coussement, K. & Van den Poel, D. (2008). Improving customer complaint management by automatic email classification using linguistic style features as predictors. *Decision Support Systems*, *44*(4), 870–882.

Dombrow, J. & Turnbull, G. (2004). Trends in real estate research, 1988–2001: What's hot and what's not. *Journal of Real Estate Finance and Economics*, *29*(1), 47–70.

Doran, J. S., Peterson, D. R., & Price, S. M. (2012). Earnings conference call content and stock price: The case of REITs. *Journal of Real Estate Finance and Economics*, *45*(2), 402–434.

Downs, D. H., Guner, N. Z., Hartzell, D. J., & Torres, M. A. (2001). Why do REIT prices change? The information content of Barrons "the ground floor." *Journal of Real Estate Finance and Economics*, *45*(1), 63–80.

Evangelopoulos, N., Ashton, T., Winson-Geideman, K., & Roulac, S. (2015). Latent semantic analysis and real estate research: Methods and applications. *Journal of Real Estate Literature*, *23*(2), 355–380.

Evangelopoulos, N., Zhang, X., & Prybutok, V. (2012). Latent semantic analysis: Five methodological recommendations. *European Journal of Information Systems*, *21*(1), 70–86.

Harrison, D. M. & Manning, C. (2008). Characteristics of recent real estate research: 2000–2006. *Journal of Real Estate Practice and Education*, *11*(2), 109–126.

Ignatow, G. & Mihalcea, R. (2017). *Text Mining: A Guidebook for the Social Sciences*. Los Angeles, CA: Sage.

Johnson, L. L., Roulac, S. E., & Followill, R. A. (1996). American Real Estate Society annual meeting paper presentations: The first decade (1985–94). *Journal of Real Estate Research*, *12*(2), 195–207.

Jud, G. D. (1996). The journal of real estate research: A ten-year review. *Journal of Real Estate Research*, *12*(2), 249–313.

Landauer, T. K. (2007). LSA as a theory of meaning. In T. Landauer, D. McNamara, S. Dennis, & W. Kintsch (Eds.), *Handbook of Latent Semantic Analysis* (pp. 467–479). Mahwah, NJ: Lawrence Erlbaum Associates.

Larsen, K. R. & Bong, C. H. (2016). A tool for addressing construct identity in literature reviews and meta-analyses. *MIS Quarterly*, *40*(3), 529–A20.

Lindholm, A. L., Gibler, K. M., & Levainen, K. I. (2006). Modeling the value-adding attributes of real estate to the wealth maximization of the firm. *Journal of Real Estate Research*, *28*(4), 445–475.

Salton, G. (1975). A vector space model for automatic indexing. *Communications of the ACM*, *18*(11), 613–620.

Wild, F. (2015). Package LSA. https://cran.r-project.org/web/packages/lsa/lsa.pdf, accessed on May 25, 2017.

Winson-Geideman, K. & Evangelopoulos, N. (2013a). Topics in real estate research 1973–2010: A latent semantic analysis. *Journal of Real Estate Literature*, *21*(1), 59–80.

Winson-Geideman, K. & Evangelopoulos, N. (2013b). Research in real estate 1973–2010: A three journal comparison. *Journal of Real Estate Literature*, *21*(2), 255–267.

Winson-Geideman, K. & Evangelopoulos, N. (2013c). Reading lists for PhD seminars in real estate. *Journal of Real Estate Practice and Education*, *16*(1), 41–82.

12 Real estate analysis case study

This chapter illustrates the analytical process described in Chapters 4 through 11 with a case study of the city of Seattle housing market. More specifically, it examines and describes two relationships: (1) crime reports and housing prices, and (2) house price changes and citizen sentiment. The focus of the case study is on providing a detailed example of the process by which a complex analysis of real estate markets is done using a variety of data sources. The results of the case study will be a by-product of the process, not necessarily scientific findings in and of themselves.

The results are presented in a narrative form, describing the meaningful choices made in the data munging and analysis process, as well as the basic results. A detailed explanation of the process and rationale for the various steps in the analytical workflow is found in the previous chapters. The step-by-step data provenance can be accessed at the book's online code repository, found at www.github.com/REAIABook/REAIABook. Requests to fix issues that users may find in the code are welcome; however, please note that in many cases the code and associated instructional documents have been prepared with transparency, not efficiency, as the guiding principle.

Seattle case study

The case study seeks to use the analytical process described in the previous chapters to explore two questions:

1. Is there a relationship between reported crimes and house values in the city of Seattle?
2. Is local sentiment related to local house price changes?

Traditional real estate data with peripheral (big) data will be utilized to explore these two questions. The traditional real estate data are home sales and property characteristics data from the King County (Washington) Tax Assessor's Office. The King County data can be freely downloaded from the Assessor's website. Peripheral data is gathered from Twitter, both from the online site as well as their application programming interface (API). The data gathering processes are described completely in the supplemental materials found on the Github site mentioned above.

After gathering the data, the general process of manage, prepare, explore, and clean is followed, as described in Chapters 6 through 9. The real estate data cover the entire county (of which Seattle is one part), all property types, and contains all forms of transactions (arm's length and not). As a result, the preparation, exploration, and cleaning processes are extensive. The code in the supplemental materials provides a step-by-step guidance through transforming field values, integrating data tables, filtering out errors, exploring data distributions, and finally labeling discordant values.

The Twitter data take the form of unstructured text and necessitates a similar process, but different individual type of data munging activities. Two sets of Twitter data are used. The first is a comprehensive set of 53,000 tweets posted by the Seattle Police Department's 51 Twitter accounts in a 6-month period in 2016. These can be collected through manual download directly from the Twitter website.[1] The first data set will help us address the first question in this case study. The second set is a comprehensive set of 113,900 tweets that mentioned "Seattle" in a 20-day period in February 2017. The second set was collected using the Twitter API, which provides access to past tweets up to 10 days old; therefore, the particular tweets used in the analysis are no longer available to the reader. However, any similar data set collected in the future should serve the purpose of helping address the second question of this case study. Both data sets underwent a data munging process that converted them to clean spreadsheet tables.

Modeling and analysis

The analytical or modeling component kicks off the case study by constructing a model of the City of Seattle housing price market for the 2016 calendar year. A hedonic price model is used for this purpose. After the filtering done in the data preparation and cleaning processes, there were 8,127 sales of single family houses (detached and townhomes) in the City of Seattle during 2016. The location of these sales is shown in Figure 12.1, with summary statistics of the major property characteristics listed in Table 12.1.

Base model

The purpose of this initial modeling exercise is to build a basic model specification that can be used in the two case study applications that follow. The concern is mostly with model robustness, efficient coefficient estimates, and the ability to control for micro-spatial influences in the market. Optimizing the out-of-sample predictive abilities is not important at this time.

The property characteristics dataset from King County is rich and contains many useful fields about each home that sold. As a result, choosing a model specification is a critical component of the model building process. Model specification is approached in two ways, first with step-wise regression and second with Bayesian Model Averaging (BMA) (Raftery, Gneiting, Balabdaoui, & Polakowski, 2005). Both techniques begin with the full set of independent variables and allow the algorithm to pare these down to the meaningful contributors explaining price variation.

This model specification process leads to a few changes to the data and specification. First, the condition, traffic noise, and view fields are reclassified so that factor levels that were too small or had an insignificant relationship with price are combined to form categories that are more meaningful. The size of a basement garage, the presence of challenging topography, and a having a lot (land parcel) of restrictive size or shapes are all found to have no measurable impact on price. These variables are dropped from the specification, and then the model is re-estimated. The initial results of the model are listed in Table 12.2. We will consider this as the "base" model.

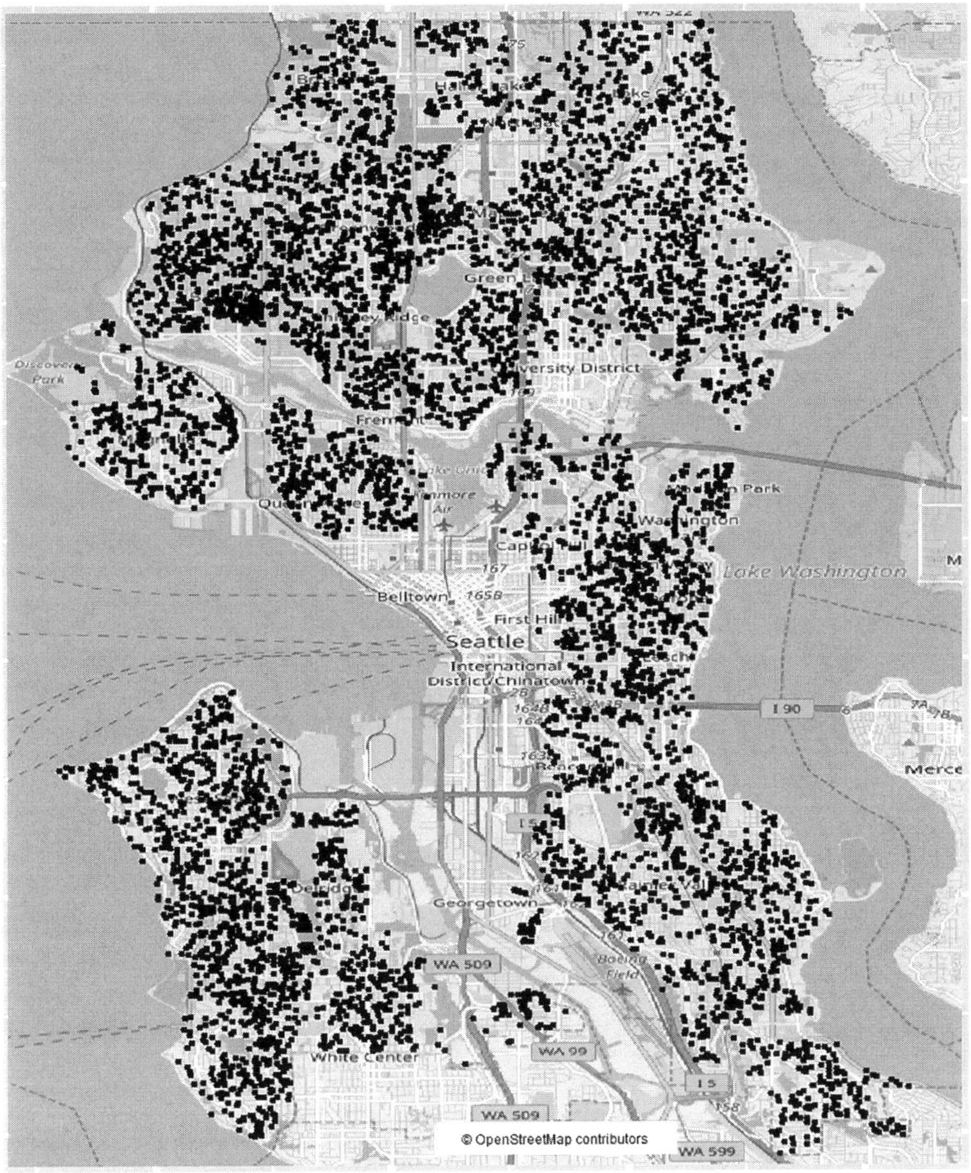

Figure 12.1 Single family housing transactions, 2016

The large majority of coefficient estimates have the expected signs, with some exceptions. For example, the coefficient on the number of bedrooms has a negative and statistically significant coefficient. This is similar to the findings of Lipscomb and Farmer (2005) and Read (2015). In particular, Read (2015, p. 155) explains occurrences like this by saying, "Number of bedrooms was found to have a negative impact on housing prices, after holding square footage constant, thereby suggesting houses with smaller bedrooms

Table 12.1 Summary statistics

Variable	*Min*	*1st Quartile*	*Median*	*Mean*	*3rd Quartile*	*Max*	*St. Dev*
Sale Price	$63,500	$475,850	$625,000	$708,228	$800,000	$9,750,000	$416,002
Lot Size	531	2500	4800	4902	6250	86707	3848
Home Size	330	1244	1590	1771	2110	9180	797
Basement Size	0	0	210	343	600	2680	403
Basement Garage Size	0	0	0	96	200	1790	168
Attached Garage Size	0	0	0	61	0	1020	143
Deck Size	0	0	0	124	200	1650	189
Stories	1	1	1.5	1.62	2	4	0.65
Building Quality	4	7	7	7.52	8	13	1.06
Bedrooms	0	2	3	3.07	4	10	0.93
Bathrooms	0	1	1.75	1.96	2.5	8	0.80
Age	0	10	66	55.05	89	116	38.42
Effective Age	0	9	62	51.04	87	116	38.36
Townhome	0	0	0	0.23	0	1	0.42
Waterfront	0	0	0	0.01	0	1	0.08
Mountain View	0	0	0	0.07	0	1	0.25
Water View	0	0	0	0.08	0	1	0.28
Discordant Value	0	0	0	0.08	0	1	0.27

Table 12.2 Base model results

Variable	*Estimate*	*Standard. Error*	*t value*	*Probability (>|t|)*
Intercept	8.490406913	0.525465268	16.15788412	7.95E-58
Townhouse	−0.101920896	0.011174443	−9.120892815	9.26E-20
Lot Size (sq ft)	−1.02E-05	1.01E-06	−10.1595524	4.19E-24
Home Size (sq ft)	0.000267764	8.73E-06	30.66486801	1.78E-195
Basement Size (sq ft)	−0.000106377	1.03E-05	−10.31576837	8.51E-25
Attached Garage Size (sq ft)	−0.000168928	2.22E-05	−7.597485443	3.36E-14
Deck size (sq ft)	9.85E-05	1.66E-05	5.916721641	3.42E-09
Building Quality	0.204892437	0.00461437	44.40312313	0
Condition = 3	0.092331334	0.026218839	3.521564506	0.000431368
Condition = 4	0.144597114	0.026548528	5.446520862	5.29E-08
Condition = 5	0.158582069	0.027755814	5.713472101	1.15E-08
Effective Age	0.002500578	0.000130316	19.18856588	2.73E-80
Baths	0.035425066	0.006272717	5.647483687	1.68E-08
Bedrooms	−0.013670654	0.004203397	−3.25228721	0.001149487
Traffic Noise: Moderate	−0.02668539	0.009266726	−2.879699715	0.003990968
Traffic Noise: Severe	−0.058541486	0.010154198	−5.765249506	8.45E-09
View of Cascades	0.062800723	0.016796975	3.738811539	0.000186165
View of City Skyline	0.257997537	0.07526905	3.427670954	0.00061181
View of Olympics	0.078245028	0.017347408	4.510473653	6.56E-06
Other View	0.019784136	0.011393782	1.736397652	0.082531545
Waterfront	0.521253357	0.043127046	12.08646106	2.41E-33
Sales Date	0.000158876	3.08E-05	5.150160958	2.66E-07
Diagnostics				
r-squared	0.66778926			
standard error	0.259759736			
f-statistic	775.8173452			
AIC	1177.034876			

tend to sell for less than those with larger bedrooms." Kennedy (2005) agrees that this is a common reason for the structural characteristics of housing units, such as the number of bedrooms, to have negative coefficients when included in regression models along with square footage.

Base model diagnostics

Next, the base model is subjected to a number of additional diagnostic tests, beginning with an examination of the variance inflation factors (VIF) for problematic levels of multicollinearity. There are no significant issues of multicollinearity in the model, as all VIFs fall below 3.[2] The highest VIF values are found in the home size and home age fields, generally due to the fact that newer homes are larger than older homes, so there is some correlation between these two fields.

In order to check for lack of variance homogeneity (i.e., heteroskedasticity), two Breusch-Pagan tests are employed: one studentized and one not studentized. The studentized test shows a significant level of heteroskedasticity in the results, while the non-studentized version does not. Next, we address heteroskedasticity by re-estimating the standard errors with White's correction. There are some small changes in significance levels once the non-constant variance is accounted for in the standard error estimates, but nothing large enough to suggest changing the base model.

Next, we check for spatial autocorrelation in the error terms of our base model. Due to the many local spatial features that impact housing prices, spatial autocorrelation is common in house price models, especially those covering a large, diverse area such as the city of Seattle. To test for spatial autocorrelation, we submit the base model residuals to a Moran's I test (Moran, 1950). As expected, a statistically significant level of spatial autocorrelation is found in the results. LaGrange Multiplier tests show that the dominant form of spatial dependence is dependence in the error terms (as opposed to dependence in the dependent variable). Next, we use a spatial error model to correct for this spatial autocorrelation. Rerunning the Moran's I test on the residuals from the spatial error model show that spatial autocorrelation has now been eliminated. The updated coefficient estimates from the base model and the spatial model are shown in Table 12.3.

The spatial model offers a much better fit, both in terms of residual standard error and AIC value. Some of the "wrong signs" change back to be consistent with economic theory, such as the positive coefficient on the Lot Size and Bedrooms variables in the spatial model. Additionally, the magnitude of a number of coefficients has changed significantly under the spatial error model, suggesting that the method to account for omitted micro-local influences on prices does improve the model.

Finally, we test for spatial heterogeneity in our results. Spatial heterogeneity would result if the coefficients in our model vary widely over space. To test for spatial heterogeneity, we estimate a geographically weighted regression (GWR) model following Fotheringham, Brunsdon, and Charlton (2002). Essentially, we estimate a separate model at each location in our data (each sale location) with observations weighted by distance to that location.

The results of the GWR test show the presence of spatial heterogeneity in the model, but it is rather small. We gauge the extent of it by examining the distribution of the coefficient estimates for each variable across the thousands of models that were estimated. For instance, coefficients for home size (logged) range from 0.45 to 0.51, with the interquartile range from 0.47 to 0.49. Overall, we found relatively high coefficient stability for home size. In general, the view and condition variables show the largest variation in coefficient

Table 12.3 Comparison of aspatial and spatial models

Variable	*Base model (OLS)*	*Spatial model (spatial error)*
Intercept	8.490406913	8.236804742
Townhouse	–0.101920896	–0.133319614
Lot Size (sq ft)	–1.02E-05	4.43E-06
Home Size (sq ft)	0.000267764	0.000208646
Basement Size (sq ft)	–0.000106377	–8.53E-05
Attached Garage Size (sq ft)	–0.000168928	–3.05E-05
Deck size (sq ft)	9.85E-05	3.76E-05
Building Quality	0.204892437	0.116454646
Condition = 3	0.092331334	0.090579999
Condition = 4	0.144597114	0.152882996
Condition = 5	0.158582069	0.18785035
Effective Age	0.002500578	0.000125226
Baths	0.035425066	0.036043063
Bedrooms	–0.013670654	0.007741093
Traffic Noise: Moderate	–0.02668539	–0.027909958
Traffic Noise: Severe	–0.058541486	–0.038257517
View of Cascades	0.062800723	0.074995831
View of City Skyline	0.257997537	0.14766849
View of Olympics	0.078245028	0.093208295
Other View	0.019784136	0.038056814
Waterfront	0.521253357	0.399345927
Sales Date	0.000158876	0.000217492

estimates across space, but are still within reasonable measures – the IQR of the GWR estimates are within the standard error estimates of the standard linear model (base model). No coefficient shows a change of sign within the IQR of the coefficient estimates.

Sensitivity tests

To check the robustness of the initial modeling exercises, the model is submitted to a number of sensitivity tests. Three major forms of sensitivity are tested: (1) sensitivity to specification; (2) sensitivity to discordant values; and (3) sensitivity to the data sample.

First, test for sensitivity to the specification by estimating four additional models with less restrictive specifications compared to the relatively restrictive (highly specified) model that is used as the base model. Three of the additional specifications are spatial error models, while one is a simple linear model. The coefficient estimates are relatively stable across the three spatial error models, which speaks to the robustness of the model specification. There is, however, a large difference in coefficient magnitudes and (occasionally) sign when comparing the simple linear model (aspatial) to the other test specifications. This again highlights the importance of controlling for micro-spatial influences on the house prices with a spatially explicit model.

Second, test for the impact of discordant values on the model results. Three different scenarios are tested for sensitivity: (1) removal of observations with discordant dependent variables (sale price); (2) removal of only those observations with univariate discordant values (no conditional or contextual outliers); and (3) removal of all suspected discordant values. As above, the coefficient estimates are found to be generally robust to the inclusion or exclusion of different types of discordant values. The coefficients for lot size and home age show a moderate change in magnitude when discordant values are removed. This finding is likely due to the fact that many of our discordant observations are very high-end outliers with uncharacteristically large lots and/or old and decrepit structures, which may have been sold for development land (as a "tear-down" in the real estate vernacular). In such cases, the sale price often represents the development potential of the site(s) and not the property as a single-family residential use.

Third, test for sensitivity to the data sample itself. Next, run 10 cross-validation (CV) tests, where the random sample is 75% of the observations and re-compute the model. Then compare the coefficient estimates from the 10 CV runs. As a result, limited changes in coefficients are found for the structural characteristics (such as age, size, number of baths, etc.) across the samples. Conversely, larger fluctuations in the coefficient estimates are found in the view and traffic noise variables. This is likely due to the existence of views and traffic noise that are prevalent in a small number of the properties in the overall sample but that are not present in some of the re-samples. This leads to high volatility in the coefficient estimates, primarily due to small sample sizes.

Code and related data sets for the analyses described above are found in the code for Chapters 5 through 10 of the online supplement. Code and data sets for the analyses that follow are found in the code for Chapter 12.

Crime and house prices

Having built a general model of the housing market in Seattle during the year 2016, next is the first of two parts of this case study that describe the process of combining peripheral data with traditional real estate data. In both cases, focus should be on the process and the potential, not the actual empirical results themselves or the theoretical validity of the underlying analysis. First, we test whether or not crime reporting, as operationalized through the City of Seattle's "Tweets by Beats" data, correlates to trends in the housing market.

The City of Seattle is divided into five geographical areas that correspond to five Seattle Police Department precincts: North, East, South, West, and Southwest. Each precinct is further divided into smaller sections, called beats, with a total of 51 beats in the entire City of Seattle. "Tweets by Beats" is a program maintained by the Seattle PD, where police dispatches feed 51 Twitter accounts organized by Beat.[3] The tweets contain the specific crime reported (based on the coded city category), as well as the time and the location of the report. A typical tweet looks as shown below:

> Beat: N2, MOTOR VEHICLE COLLISION at N 122 ST/PALATINE AV at 2/7/2017 3:14 PM, Call# 19000011111.

The process for compiling these data is described in detail in the supplemental materials to this book on the Github site. The original home sales data from the Tax Assessor's Office are recorded at the individual property level, whereas the crime data are recorded at the Beat level, with 51 levels in the entire Seattle area. We are consolidating the two data recording levels by relating the price of each property to localized crime. "Localized" here is defined as occurring within ¼ mile or 400 meters of a property. Impacts are measured based on the number of reported incidences occurring one year prior to the sale date, with the incident counts measured in 100s. There are 54 different major categories of crimes, the top 15 of which account for 90% of all reported incidents. These 54 categories are grouped into five major types: (1) violent crime; (2) property crime; (3) behavioral crime; (4) traffic crime; and (5) other crime. Figure 12.2 shows the breakdown of the incidents by major type for the period of data collection in 2016.

Next, we add the counts of localized crime by type to the hedonic price model developed in the previous section, first to the aspatial specification and then to the spatial error model. Table 12.4 shows the coefficient estimates for the crime variables from both models. Estimates show the impact on home price (in percentage) for every 100 of the various crime types occurring within one year and ¼ mile of the sale. In the aspatial specification,

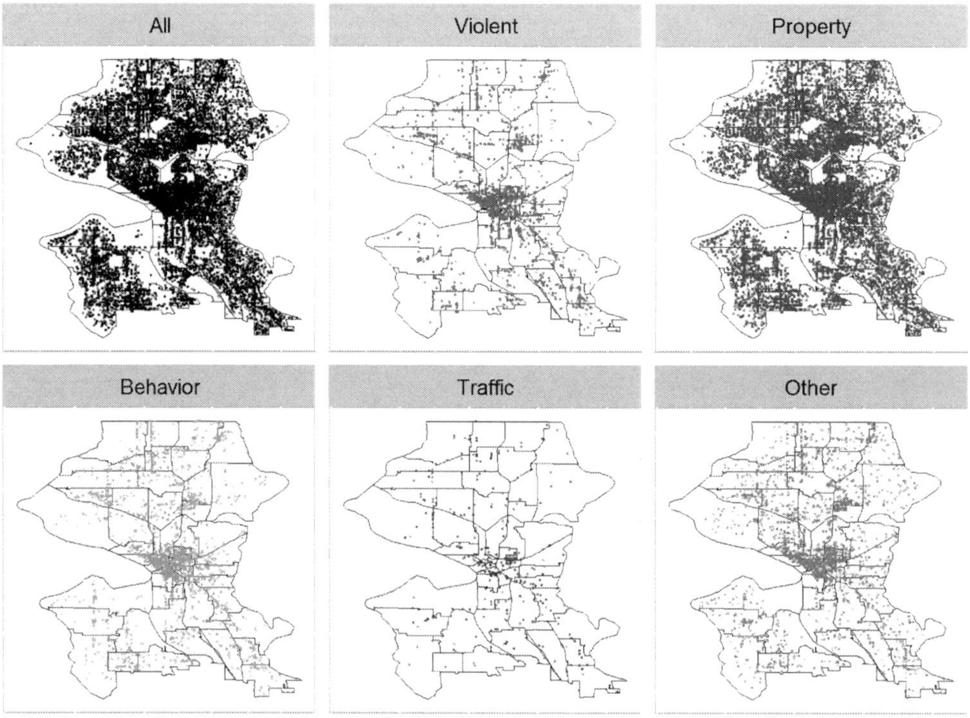

Figure 12.2 Crime by major type

Table 12.4 Crime types and house prices

Crime Type (100 within 1/4 mile)	*Linear Model*	*Spatial Error Model*
Violent Crime	***–0.21347***	0.03949
Property Crime	***0.07336***	0.00688
Traffic Crime	–0.23854	0.41565
Behavior Crime	***–0.16492***	0.01829
Other Crime	***0.17703***	–0.01851

Note: Bold italic indicates significant at 0.05 level

only the traffic crime has no significant impact on prices. Violent and behavior crime show negative impact, while property crime and others show positive impacts. The positive impact of property crime is likely due to the fact that property crime is more prevalent in areas with higher value property, whether real estate, cars, or household goods. However, once the inherent local spatial structure is accounted for with the spatial error model, the impact from the crime variables, across all crime types, disappears.

The above analysis suggests that the impacts of crime on housing prices may be subsumed in larger or more prevalent localized amenities or disamenities. For instance, past crime may have led to a decrease in home maintenance in the area and/or a decrease of the quality of public goods, and both factors that may have decreased prices across larger neighborhood areas. Alternatively, the 400-meter (¼ mile) boundary may not be the correct scale at which to measure the effect. The real estate researcher can easily test different distance boundaries as a robustness check.

If the impact of crime is part of a larger neighborhood effect, then perhaps crime heavy areas may also be suffering from decreased price appreciation, as compared to the larger city. To test this, we compare the rates of price appreciation in 2016 with the number of crimes per square mile for each of the 51 police Beats. To ensure robust estimates of price movement over 2016, we limit the Beats included in this analysis to those with more than 100 residential sales in the year 2016. In the plot below, crimes per square mile can be compared to the change in prices in each of the selected beats from the first quarter to the fourth quarter.

As Figure 12.3 shows, there is no consistent relationship between the number of reported crimes and the house price appreciation within each police Beat. For some

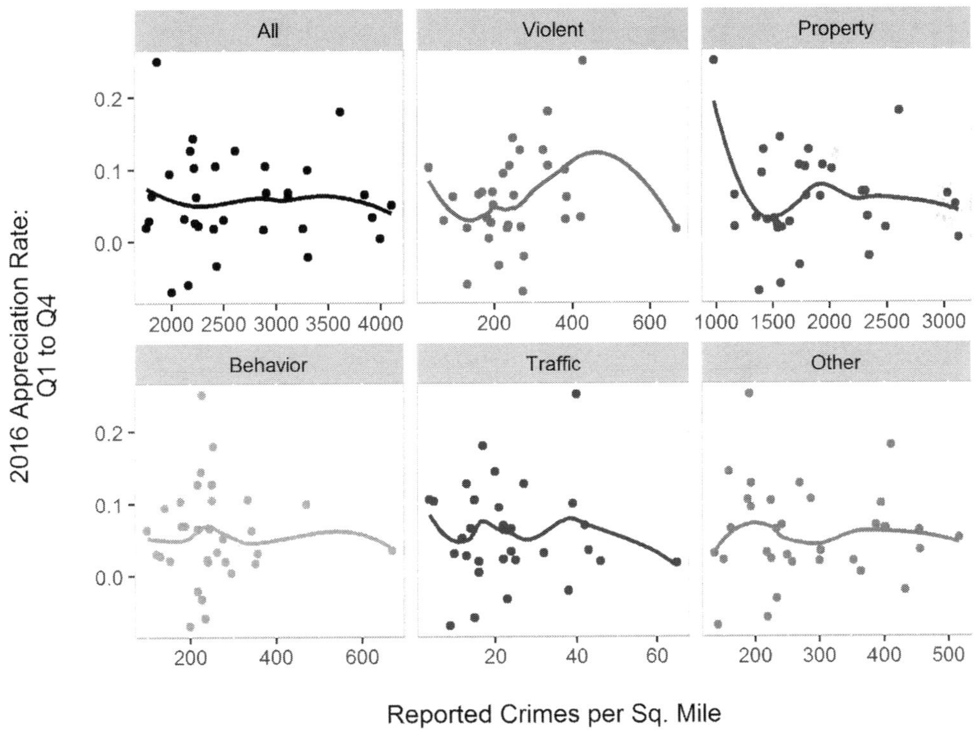

Figure 12.3 Crimes versus price appreciation

crime types, the relationship is clearly negative, but for others it is generally flat. A number of different factors are at play that may explain this set of results. Property depreciation may indicate a continuing effect of various types of crime on home prices. A flat appreciation rate may indicate that the effect of crime is fully absorbed by the market and home prices have stabilized at relatively higher (for low-crime areas) or relatively lower levels (for high-crime areas). Property appreciation may be driven by the fact that many of the lower priced (and often high-crime) areas of the city of Seattle are currently experiencing significant gentrification pressures, causing prices to rise quickly in these areas.

Sentiment and house prices

The second component of the case study investigates the relationship between citizen sentiment and the 2016 house price movements in the City of Seattle. Here we define citizen sentiment as the sentiment of Twitter users posting within the city limits. After collecting a comprehensive set of 113,900 tweets that mentioned "Seattle" in a 20-day period in February 2017 using the Twitter API and the R library Twitter,[4] we identify 6,754 tweets, or about 6 percent, that contain geo-location data. These tweets are then scored for sentiment.

Rather than focusing on facts, such as the crime incidents examined in the previous section, sentiment analysis focuses on subjective opinion. The particular version of sentiment analysis we employ in this case study examines two types of opinion: positive and negative. The R code that generates sentiment scoring is provided in the supplemental web site. The code uses Bing Liu's lexicon of 2,006 positive and 4,784 negative words to compute sentiment score. Examples of positive words include *appreciate* and *beautiful.* Examples of negative words include *absurd* and *barbaric.* The sentiment score of a certain tweet is a simple total count of positive words in the tweet minus the total count of negative words in the tweet. For example, the presence of words *appreciate* and *beautiful* in a tweet would result in a sentiment score of +2, the presence of words *absurd* and *barbaric* would result in a sentiment score of –2, while the presence of all four words – or none of the four – would result in a sentiment score of 0. For more details on this type of sentiment scoring, see Hu and Liu (2004) or Liu (2010).

We begin the incorporation of the sentiment data into our data set by converting the geocoded and sentiment-scored tweets into a two-dimensional 'surface' showing the average sentiment at each area of the city. The point data are converted to an areal surface through an inverse distance weighted (IDW) method. First, we convert the city into ¼ mile grid points, and then estimate the sentiment at each point using the IDW technique. Figure 12.4 below shows the average "sentiment" across the city.

Next, we use a geographically weighted regression (GWR) model to estimate the local house price appreciation at each of the ¼ mile grid points. Figure 12.5 shows the localized change in house prices at each of the grid points, where we define localized house price change as the change from the first quarter to the fourth quarter of 2016.

To test if there is any relationship between the two, we plot the local sentiment versus the local house price movements. As seen in Figure 12.6, there is little discernable relationship between the two variables. There are several possible reasons for this – only a small sample of tweets are geolocated, and those may not be from homeowners, or even local residents.

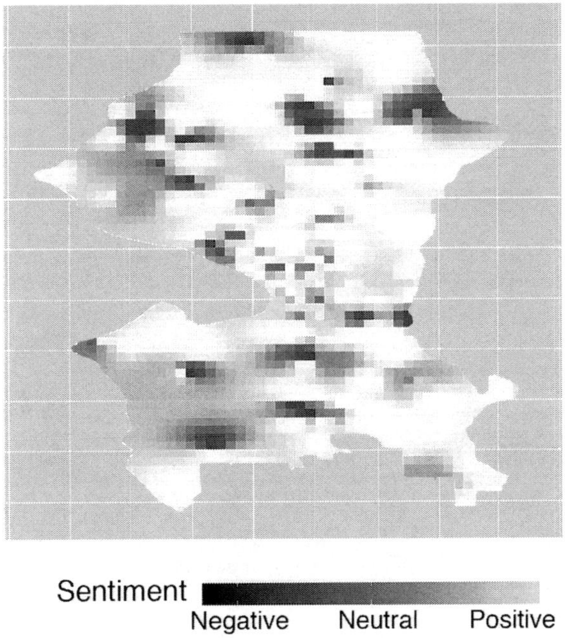

Figure 12.4 Sentiment in Seattle

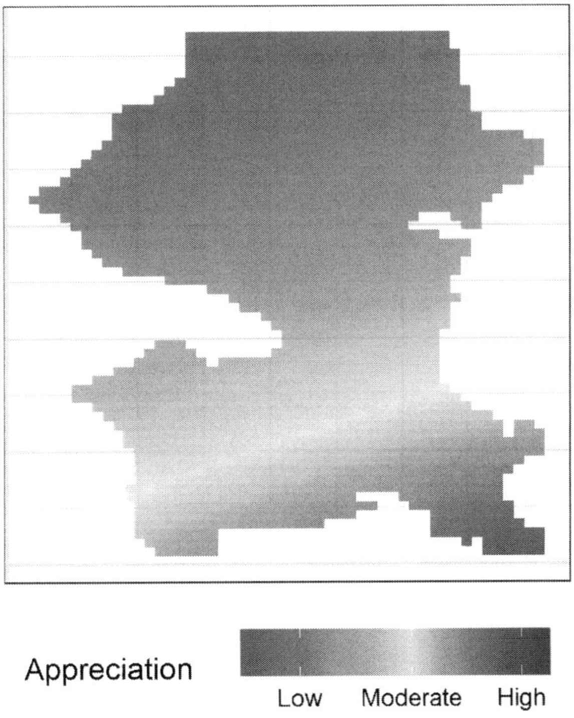

Figure 12.5 2016 Price appreciation in Seattle

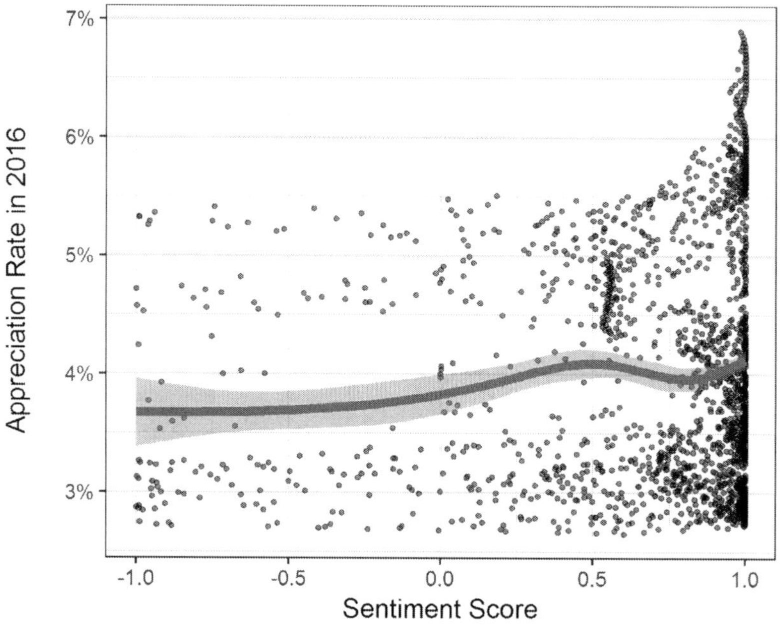

Figure 12.6 Price movement versus sentiment

Conclusion

This chapter has provided an overview of two complex real estate market analyses that leverage both traditional real estate data and peripheral (big) data. The description above focuses on the underlying process used to perform such analyses. More details about how each individual step in the analytical workflow (the data provenance) was performed, from data collection through model estimation, can be found in the supplemental materials hosted on the Github site mentioned at the start of this chapter.

In our examination of the impact of crime on house prices and the relationship between sentiment and house price movement, only weak, if any, relationships are found. These findings, or lack thereof, are not surprising given the focus on process and illustration, as opposed to theory building and hypothesis testing. Further, the findings in our case study do not deny the possible relationship between crime and house prices; rather, they indicate that the particular approach used here may not adequately measure or express the relationship(s), if they exist. For example, we attempted to account for price changes between 2016Q1 and 2016Q4 using crime incidents or citizen's sentiment; a panel data set that considers a longer period of, say, 5 or 10 years, might be more appropriate for capturing the relationship between property appreciation and crime facts. Another point to consider is that, perhaps property prices do not depend on the local positive or negative sentiment. While sentiment may not, then, be a factor that determines prices, it can still be an interesting piece of information for the future buyer, who may want to buy property in a neighborhood with positive mood, especially if that does not command a higher price. Finally, the results presented in this case study also highlight the fact that, in most cases, data itself is not enough. Careful thought must be given to the design of a meaningful research question, project, or product, and then collecting data and developing an analytical workflow that can best address domain-specific issues in real estate (such as spatial heterogeneity, spatial autocorrelation, etc.).

Notes

1 https://twitter.com, manual search for handles @SeattlePDN1, @SeattlePDN2, and so forth.
2 Standard cutoff for VIFs is 10. See, e.g., Kutner, Nachtsheim, and Neter (2004, p. 409).
3 www.seattle.gov/police/information-and-data/tweets-by-beat, accessed on 25 May 2017.
4 https://cran.r-project.org/web/packages/twitteR/twitteR.pdf, accessed on 25 May 2017.

References

Fotheringham, A. S., Brunsdon, C., & Charlton, M. (2002). *Geographically Weighted Regression: The Analysis of Spatially Varying Relationships*. Chichester: Wiley.

Hu, M. & Liu, B. (2004). Mining and Summarizing Customer Reviews. *Proceedings of the ACM SIGKDD International Conference on Knowledge Discovery and Data Mining (KDD-2004)*, Seattle, Washington, USA, August 22–25, 2004.

Kennedy, P. E. (2005). Oh no! I got the wrong sign! What should I do? *The Journal of Economic Education*, *36*(1), 77–92.

Kutner, M. H., Nachtsheim, C. J., & Neter, J. (2004). *Applied Linear Regression Models*. New York: McGraw-Hill/Irwin, 4th edition.

Lipscomb, C. A. & Farmer, M. C. (2005). Household diversity and market segmentation within a single neighborhood. *Annals of Regional Science*, *39*(4), 791–810.

Liu, B. (2010). Sentiment analysis and subjectivity. In N. Indurkhya, & F. J. Damerau (Eds.), *Handbook of Natural Language Processing* (pp. 627–666). Boca Raton, FL: Chapman & Hall/CRC, 2nd edition.

Moran, P. A. P. (1950). Notes on continuous stochastic phenomena. *Biometrika*, *37*(1–2), 17–33.

Raftery, A., Gneiting, T., Balabdaoui, F., & Polakowski, M. (2005). Using Bayesian model averaging to calibrate forecast ensembles. *Monthly Weather Review*, *133*, 1155–1174.

Read, D. C. (2015). The impact of an adequate public facilities ordinance on the sale price of single-family housing in Cabarrus County, North Carolina. *Housing and Society*, *42*(2), 148–161.

Part IV

Legal and future

13 The legalities of big data

Lucy Cradduck

This chapter's discussion is undertaken from the perspective that, as a matter of good governance, it is necessary to ensure that legal compliance obligations are met. Importantly, the starting point for properly managing any legal risk, which is any risk arising from non-compliance with a legal obligation, is to understand that legal obligation. To do this necessitates identifying legal rights and the party/ies in which these vest; and legal obligations and upon whom these are imposed. The primary "areas" of law considered in this chapter are those arising under specific laws; more commonly for big data these are privacy and data security as well as aspects of intellectual property.

The reality of online life, however, is that our engagements need to be facilitated by non-government entities. Relevantly, this includes both your ISP and, in the case of big data, the collector/keeper of big data. These interactions, while also governed by formal laws, more commonly, particularly as relates to access and use, are primarily regulated by the terms of the use agreement between the keeper and the party/ies seeking to access or utilize that data. The chapter therefore concludes with a consideration of such contractual relationships and the obligations more commonly imposed.

Introduction

Irrespective of the source of the data, or whether it is public or private, there are three perspectives that are relevant. The first is that of the creator of the data. In the case of social media sites, this includes you, me, and the other members in respect to the information that we post or otherwise upload. In the property specific context, "creators" also would include the owner of the information posted on the site where this information is posted and/or created by an agent on their behalf. The second is that of the "site owners," that is, the entities that own the sites on which the information is posted. These may be the same as the data owners but may be separate legal entities. The third is of the "users," that is the other parties who have access to the sites and who may wish to access and/or use the information/big data available on those sites.

In many respects the concerns regarding big data are the same as for any data set. The significant difference being, as earlier chapters have discussed, is that big data is, well, *big*, and the bigger the data, the harder it falls. The issues arising in respect of big data are to a large extent the same as those that are relevant to the handling of any data, with two notable exceptions. The first is that big data is constantly being created, and therefore the need to ensure compliance with obligations regarding accuracy, for example, as contained within privacy laws, is constant. It is not a matter of "set and forget." The second is that if there is a system or process failure, the resulting breach also will be big. Where the breach is of contractual use obligations, the resulting penalty (in the terms of financial damages) is also likely to be big.

Particular concerns arise for business when the big data concerned is shared by government agencies with the business. This sharing usually occurs pursuant to the terms of a contractual agreement, which more commonly impose upon the non-government party obligations regarding data storage, security, and sharing that may not otherwise be imposed upon that party by the law itself: for example, where a business is not of a size to require it to comply with certain privacy obligations but there is a term in the tender contract by which the successful tender agrees to be bound by those very obligations. Such obligations, and the risks of breaching those obligations, therefore, are ones for which the business' systems and personnel often are unprepared and may not be compliant.

Breaches of data obligations are often reported in the media, which can use emotive language to describe the breach and the consequences for the parties adversely affected. While clearly this would have a negative impact for any business, for the purpose of this text, the risks discussed here are considered from the perspective of breaches of legal obligations only for which there are legal remedies or sanctions available, rather than from the perspective of any "moral obligation" (McDermott, 2017).

While risk avoidance is the ideal, this may not be able to be guaranteed. Risk minimization, which includes ensuring proper education of employees, directors, and other officer bearers; establishing appropriate access limitations; and adopting best practice for data handling, therefore should be the focus. Preventive policy and procedures, while possibly costly to implement, are always preferable to finding yourself in the position of trying to fix the unfixable. Regretfully many businesses are reluctant to spend money proactively and so do not invest in preventive action only to later find their frugal behavior was a false economy.

Uses that may cause particular concern for big data include automated decision-making and profiling functionalities; decisions (or lack of decisions) regarding who is (should be) authorized to access data; and unthinking file sharing. In addition to such breaches, issues also will arise in respect of deliberate actions: for example, regarding health or medical related records or financial information. In the property context, the data that may be at risk can include rental and sale data, profit and loss statements and projections, tenants' turnover information, and the personal contact details of tenants, employees, and contact officers.

The two primary threats/challenges to any data set are those that arise from privacy and security breaches (Morabito, 2015). These potentially can result in significant adverse impacts for a business as well as for any individual concerned. As a consequence of any breach of either privacy or security obligations, the related data could be used to facilitate identity theft for any individual/s whose data was affected by the breach. Where the information concerned is part of a big data dataset, the problem is magnified due to the volume of information and number of parties affected as an instance of privacy or security breach may affect the data of many individuals. Separate from any repercussion under related laws by the relevant regulator, such a breach may leave a business vulnerable to multiple (or in some countries class) court actions.

Legal liability clearly will arise in respect to breaches of privacy and security obligations/laws; however, even when there is no breach of privacy or security, the mere use of data of which you are not the owner may breach other laws, even if the conduct was not deliberate. Information is an asset and the ownership and permitted uses of data generally falls within the realm of intellectual property laws, which can provide significant remedies, such as damages or injunctions, to a party whose rights are infringed. The intellectual property rights most relevant for any data and this chapter are copyright and the associated moral rights (not the same as moral obligations) arising under copyright regimes.

Separately from these matters, and beyond the scope of this chapter to consider, businesses need to take steps to ensure the currency, completeness, and accuracy of the data they are using, as well as to ensure that it is gathered from reliable sources (Smith, 2016). Data that is incomplete or inaccurate can have fatal consequences for a property's use, value, or potential. The need to properly value big data and ensure its proper management, protection, and use therefore cannot be underestimated.

The balance of this chapter considers the legal obligations and the associated risks in the order of privacy and data security; and then intellectual property, focusing on copyright and moral rights. To provide context reference will be made to specific legislative provisions, i.e., sections of laws from the U.K., EU, U.S., and Australia, where relevant, or to international treaties. More common international treaties also are listed. Importantly, the information contained in this chapter is not a substitute for your engagement with the specific laws of the country/ies or jurisdictions in which your business operates. The chapter concludes with a consideration of contractual access and use matters.

Privacy and data security

Various domestic laws, standards, and/or regulations govern the protection of data. Relevant international treaties/documents include:

- *Universal Declaration of Human Rights*
- *International Covenant on Civil and Political Rights*
- *OECD Privacy Guidelines 2013*
- *The Lisbon Treaty*
- *Directive 95/46/EC* (Data Protection Directive) (1995)
- *Data Protection Act 1998* (U.K.)

Ensuring privacy and security obligations are met is a challenge for good governance (Almeida, Doneda, & Monteiro, 2015) and a matter of international attention (Ponemon Institute, 2011). These obligations often are linked together. However, as many countries' domestic legal regimes deal with these matters under separate laws, this chapter, as other authors have done, will approach privacy and security as being separate issues (Raja, 2016).

In this context the term "privacy" will be used to refer to the "use and governance of personal data," while the term "data security" refers to "protecting data from malicious attacks and the misuse of stolen data for profit" (Jain, Gyanchandani, & Khare, 2016). Therefore, privacy breaches relate to the failure to comply with laws that impose obligations regarding individuals' data; and security breaches relate more generally to attacks on and misuse of corporate data, which may also include information regarding individuals.

Resulting problems include the concern that the information could be used to facilitate identity theft (Cradduck & McCullagh, 2007). The need to protect privacy and ensure security are recognized as foundational elements of the international and cross-border sharing of data and information (OECD, 2013). Preventative measures, which may assist in protecting personal data and ensuring data security, include firewalls, encryption technologies and authentication systems (Shah, Ahmed, & Soomro, 2016), de-identification of data and/or access controls (Rahmani, Amine, & Hamou, 2016). Separately, and outside of what is considered here, due to the role that data now plays in economic activities, some uses of big data may be in conflict with obligations under competition or anti-monopoly laws (for further discussion, see Autorité de la concurrence and Bundeskartellant, 2016).

Privacy

Privacy concerns need to be considered in the context of the big data lifecycle of generation, storage, and processing (Jain et al., 2016) and must be balanced against the desire/need to use the data itself (Rahmani et al., 2016). From the perspective that privacy relates to the individual, it is necessary to understand to what data privacy protections attach.

Most countries have some form of privacy protection.[1] Specific laws can be very technical in nature (Svantesson, 2013) and tend to provide both a level of protection to one party at the same time as imposing obligations on another party regarding defined information (Kobus et al, 2015). This legislated protection can be found in laws specifically badged as privacy or data protection laws, in provisions contained within other laws or privacy principles, or within the country's constitution. In many countries this protection is found in one country-specific law but in other countries individual states or territories within a country will also have their own laws or principles (Australia and Germany, for example). EU member countries have the obligation to implement the specific EU privacy and data protection treaties into their domestic laws.

Privacy laws regulate behavior within a set country's jurisdictional limits; however, many countries now also seek to extend their laws extraterritorially to data, parties, or acts located or occurring beyond their borders (Svantesson, 2013). In jurisdictions where the protections are not as strong as in other jurisdictions, bi-lateral treaties are used to impose higher obligations on entities in the second jurisdiction in their dealings with data from the first jurisdiction (for example, see the EU–U.S. Privacy Shield).[2]

Privacy obligations are commonly linked to the privacy of personal information or personal data, with attaching protections and rights granted to individuals. The corresponding obligations regarding that data can be imposed upon government entities only and/or can extend to private businesses and/or can be imposed only of business of a certain size.

In the UK, the *Data Protection Act 1998*, Section 1 defines personal data as:

> data which relate to a living individual who can be identified –
>
> (a) from those data, or
> (b) from those data and other information which is in the possession of, or is likely to come into the possession of, the data controller,
>
> and includes any expression of opinion about the individual and any indication of the intentions of the data controller or any other person in respect of the individual;

In the United States, the equivalent provision in the (Privacy Act, 1974), 5 U.S.C. § 552a (a)(4), defines a "record" as:

> any item, collection, or grouping of information about an individual that is maintained by an agency, including, but not limited to, his education, financial transactions,

medical history, and criminal or employment history and that contains his name, or the identifying number, symbol, or other identifying particular as signed to the individual, such as a finger or voice print or a photograph.

In Australia, Section 6 of the (*Privacy Act 1988*) (Cth) extends the definition of personal information to include inaccurate/false information:

> personal information means information or an opinion about an identified individual, or an individual who is reasonably identifiable:
>
> (a) whether the information or opinion is true or not; and
> (b) whether the information or opinion is recorded in a material form or not.

On a physical level, personal data would include bank account numbers, telephone numbers, photographs, tattoos, social security numbers, and even motor vehicle registration numbers. The latter now resulting in the obscuring of people's faces as well as car number plates on the evening news. Other examples of what could be deemed to be personal information or personal data include "geographical location, academic and professional background, membership in groups, interests, preferences, [and] opinions" (Morabito, 2015, p. 35).

However, while many countries link obligations to what is defined to be personal information or personal data this does not mean that there is a personal right to privacy *per se*. Rather, what these laws establish is a system under which, assuming any data falls within the prescribed definition, gives the individual rights about the collection and management of that data. In countries where there is in fact a right to privacy *per se*, this commonly is dealt with under a specific law, or forms part of the country's constitution, or falls within its negligence laws. Privacy obligations are usually imposed as to how data is collected, stored, used, and shared; as well as imposing obligations to ensure that what is captured is accurate and providing a mechanism that requires the data holder to correct the data if the person to which it relates requests this.

While laws badged as privacy laws generally apply only to the information or data of individuals, in some countries these protections also are extended to apply to other legal entities. In other countries these other entities are only protected under the privacy law if they are of a certain size; and in yet other countries the laws only apply to information collected by government departments or other government entities. Absent a specific privacy law, if personal information is inappropriately captured, handled, or used, then the person affected is not without all legal redress as they may be able to take action under other laws, for example, those relevant to consumer protection or negligence.

Privacy law usually will also provide consequences for a breach of the rights it establishes. These remedies can include civil penalties, such as an injunction to restrain a repeat of the breach, and fines. Criminal penalties also may arise, which can include fines, and custodial sentences. The severity of any penalty can be linked to whether the privacy breach was willful or accidental; and whether or not the current breach is a first offense. Some exceptions to what would otherwise be a breach may apply. For example, a law may specifically permit, or exclude from application of a penalty provision, a disclosure that is made for a defined purpose. These purposes can include obtaining legal advice, if directed by a court order, to protect life or in an emergency, or as authorized by another law.

Data security

The ease of collection, use, and dissemination of data facilitated by the underlying technologies enabling and supporting big data arguably has outpaced the capacity of the various jurisdictions' legal regimes to regulate what is otherwise inappropriate conduct. This inability has been exacerbated by the enabling places and spaces of the Internet and World Wide Web. Security issues arising for a business are either direct (i.e., their own data has been compromised) or indirect (i.e., the data they hold of others has been compromised). In this respect there is, as has been noted, an overlap with privacy obligations.

Regulatory approaches to promoting security attempt this variously by imposing security obligations on the data holder; and/or providing remedies, including compensation, to the party whose data was compromised; and/or imposing obligations to report the breaches to the regulator and/or publicly, as well as notifying the affected party (Ishii & Komukai, 2016). In some countries, data security breaches are dealt with under general negligence or tort liability laws; while in others, the remedy/ies is found within the specific data protection law.

Data security laws, however named, impose obligations regarding both personal and other data and how this must be stored, and shared; as well as imposing reporting obligations if security is breached. These laws can influence where data is stored and impose both civil and criminal obligations if the law is breached. Cloud computing, while a cost effective tool for business use, raises another level of security concerns (Sodiya & Adegbuyi, 2016). Once in the cloud, the data is outside of a business' control and, dependent upon where the cloud provider is located, the cloud provider may not be subject the same level of legal obligations as the businesses using it (Jain et al., 2016).

How effective domestic laws are in meeting the challenges facing digital security issues generally depend upon the country itself and its chosen method of regulation (Ishii & Komukai, 2016; Lodder, 2013). The potential risks to privacy and data security can be seen in the four minor distributed denial of service attacks on the Australian Bureau of Statistics website on the night of the 2016 Australian Census, which resulted in the ABS shutting its site (Dudley-Nicholson & Bickers, 2016).

The impact of data security breaches can be seen in public hacking attacks, examples of which are seen in the attacks on public companies, and public figures such as Colin Powell and John Podesta (Franceschi-Bicchierai, 2016). These enabled hackers to gain access to corporate databases through web-based email accounts. While such breaches are the work of external parties, their very fact may trigger reporting obligations and liability under related laws on the hacked party. There is therefore a need for improved security generally (Gregory & Glance, 2013).

Intellectual property rights

Relevant international treaties include:

- *Berne Convention for the Protection of Literary and Artistic Works*
- *WIPO Copyright Treaty*
- *TRIPS Agreement*

Most countries have laws that deal with what are collectively referred to as intellectual property rights (IPR). Many countries also are signatories to one or more of the various international treaties, which require them to implement a minimum level of protections for IPRs within their own country; however, the rights granted are very much country specific and not necessarily consistent with other jurisdictions.[3]

Related rights and obligations will affect data whether in print or electronic-only format. IPR generally encompass matters such as copyright and moral rights, patents, trademarks, trade secrets, and/or confidential information. Some IPR laws protect the data in its actual form (i.e., copyright law) whereas others protect the idea (i.e., patent law). Some also require the recognition of IPR created in other countries.

In many countries IPR is solely found within specific written law. In some common law countries, for example the U.K., Australia, New Zealand, Canada, and the U.S., some rights and obligations arise under common law (case law) rather than a written law (legislation) (i.e., passing off). Some rights only arise upon formal registration with the relevant regulatory authority (i.e., trademarks and patents), while others arise through the mere creation of the related document (i.e., copyright). IPR can be owned by the person who originally created it (referred to as the author or inventor), their employer, or by another party. It also (generally) can be sold, assigned, licensed, transferred, or bequeathed. Irrespective of which right applies, protecting that right usually requires the party alleging that their right has been infringed to bring a civil court action to enforce and protect it.

The most commonly occurring IPR, and it is suggested the rights most relevant to big data, are those of copyright and moral rights. While in many countries these rights will exist immediately upon a work being created, some countries, particularly regarding copyright protections, will require the work to be registered to gain protection. In other countries, registration may be available on a voluntary but not obligatory basis, which may assist with resolving ownership disputes.

Identifying who owns data is a key concern particularly when what appears to be public data may in fact be private as, in order to deal with the IPR legally, it is the legal owner who needs to give permission for the proposed use. While this is in particular essential in any dealings with IPR, it also is necessary in order to ensure compliance with privacy obligations regarding the data's use or disclosure (Morabito, 2015).

Copyright

Copyright law protects the tangible form of *works* by reserving certain rights to the owner. Protection is granted to the material form in which ideas or information is expressed. To attract copyright's protection, the *work* must be original – meaning it must not be a copy of another work. Protection may be given to the written word (in printed or electronic form), artist works, musical scores, cinematography, sound records, television broadcasts, performance works, and architectural drawings.

For big data, the relevant exclusive rights would include the rights of reproduction, publication, adaptation, dissemination by any other means, sale, and licensing. Where copyright exists this can last for up to the life of the author (creator of the work) plus 70 years. Copyright rights (and obligations) are tied to the ability (or not) to engage with a specific country's law. Issues may arise for big data in the circumstances of how the data is collected and in the management of the data set created.

The first consideration is whether the data is recognized as being a *work* under the law. Some countries extend copyright protection to mere facts, databases or public information but many others do not. The second consideration is whether any action breaches that law. Infringement occurs by a party undertaking any action reserved to the owner without authorization or pursuant to any exception.

The third consideration is whether the law provides any defense or exception to the action for what would otherwise be a breach. Exceptions, depending on the country, can include fair use; fair dealing for the purpose of study, research, satire, or reporting the news; educational purposes; limited private backup; and for the purpose of assisting a person with a disability.

Copyright protection is jurisdictionally based in that there also will be a requirement for the work to have a connection with a particular country in order for the owner to enforce a right under that country's law. This usually requires the work be first published within the country or its dominions, or the author lives there or is a citizen of that country. In order to enforce copyright rights, the party bringing the action also must have a sufficient connection with the work. Other parties who would be able to bring an action would be authorized licensees. In a business sense, this includes franchisees where they are authorized by their franchise agreement to use data sets.

Moral Rights

Moral rights, importantly, have nothing to do with morality. Drawing upon the French *droit moral*, these rights, linked as they are to the economic right of copyright, provides protection for the non-economic rights associated with copyright works. Moral rights exist in respect of a work to which copyright protections attach. Moral rights are personal to the original author or authors of a work, if an individual, and last for as long as copyright in the work exists.

Unlike copyright, moral rights cannot be transferred or assigned, however, after the death of the author these may be able to be exercised by their estate. If copyright in the work is assigned, or the work is sold (i.e., a painting), or otherwise is deemed to vest in another (i.e., the author's employer), the original author still retains the moral rights in that work. The three generally recognized moral rights are:

1 Attribution as the author of a work
2 Preventing false attribution to another
3 Integrity

Remedies for infringement include damages, injunctions, declarations, public apology, and removal of false attribution or derogatory treatment. An author's moral rights cannot be waived; however, specifically worded consent may be able to be given to what would otherwise be infringing conduct. For example, an artist can consent to the reproduction of a painting that does not identify them as the painter.

Whether an act is an infringing act will depend upon any limitations contained with the particular law, for example, as regards to whether attribution was reasonable in the circumstances. Further, as regards false attribution, whether the act was done knowingly or not may be relevant to whether there is an infringement and also the appropriate penalty if so. The law may provide defenses if, as relevant, specific procedures are followed.

As regards to any moral rights obligations for big data, specific attention should be given to the moral rights of any data set created and consents obtained from all authors regarding proposed dealings with the data where relevant. There is often an impulse to try to obtain blanket consent within an employment contract; however, depending on the jurisdiction, these are likely to be ineffectual.

Contractual access and use considerations

As a starting point it is important to remember that a contract is an agreement (usually) between two parties, which regulates some behavior. Very typically in property related matters these agreements include contracts between buyers and sellers of land. Property related contracts also include agreements between landlords and tenants. Although these may be badged a "lease" or a "license to use," such terms do not make the documents some magical legal creatures as these are merely a particular type of contract. While many jurisdictions have specific laws relevant to such property transaction, the primary document for any transaction is, well, the document itself. For example, a land purchase contract sets out the actual terms of the agreement between the two parties as to the property sold, the price to be paid, what is included in the sale, and the date on which the buyer will become the owner.

Similarly, for social media and other such sites, your engagement with that site is primarily governed by your contract with the site owner. You will have agreed the terms of your contract with the site owner when you became a member of that site, which (usually) will include a provision that enables the terms and conditions to be updated by the site owner upon giving required notice to you. Many sites also have separate (and more onerous) agreements for developers and commercial entities. Further, while the terms of use may be the same or very similar for every user, your access and use is governed by *your* agreement, that is, an individual user's terms and conditions most likely will differ from that of an individual who runs a business accessing and using data. It is therefore important that before you seek to upload or use any data from any site you understand your contractual rights and obligations. Experience shows, however, that most people merely click "I Agree" without thought, with only a few reading all the terms before doing so.

It is common to distinguish between private and public information. However, even if a user has his or her profile set to "public" so that everyone may see their full profile and everything they post, this does not mean that anyone can just use that data. First and foremost, the laws mentioned above, particularly regarding IPR, will regulate proposed uses. In some countries, there will also be defamation laws that would impact upon certain uses. Secondly and more importantly, your access to and proposed use of that data will be regulated by the terms of your agreement with the site and the purpose of your use. Even when data has been sold to you, it is likely that the terms of the sale agreement will contain conditions that regulate or restrict your use of that data. In some instances the terms and conditions may become binding on you by you having merely accessed the site.

Agreements are likely to regulate separately the manner of gathering the big data, how it is permitted to be stored, and the potential uses. Depending on the site, big data may be gathered by manual download, web scraping, and/or web crawling; however, not all these methods of collection may be permitted by the site owner. Once collected, potential uses for the data include those for research, comment, or commercial activities. Research and analysis activities required, for example, for books such as this are likely to be acceptable; however, depending on how the research is conducted and the data sources accessed, the agreement may require you to obtain the consent of the member, user, or owner of the data.[4] Selling or reproducing the data set analyzed, or other commercial uses is likely to be prohibited, as would be the selling of user IDs, or otherwise breaching any other legal obligation.

Importantly, and unlike laws, which are required to go through (often lengthy) public processes for implementation and/or change, terms of access to big data can be more easily changed. In many jurisdictions this occurs by the site owner updating the standard terms and conditions on the site and then notifying existing users of the date in the future that the new terms will apply to them, with new users signing up to the displayed terms and conditions. Therefore, before you seek to access, collect, and/or use any big data set, it is important to check the current terms of your agreement (and save these as a record) so that you do not inadvertently breach these terms. Whether a commercial agreement or not, most agreements will include provisions that impose consequences (including financial penalties) for any breach of the agreement.

Conclusion

Reviewing business practices is essential to ensure the proper management of big data (Couldry & Powell, 2014) and educating employees is a crucial part of big data use and management. Absent the relevant skills, employees are not going to know what they are doing with the data and whether that use is permitted or not; nor whether that use is in breach of some legal or other obligation (Morabito, 2015). Big data requires a different skill set and knowledge in order to be able to engage with it properly (Morabito, 2015).

You should, therefore, engage with big data in a conscious manner. That is, in your dealings with big data (both that belonging to you and to others), your focus should be on what you are doing, where you are doing it, why you are doing it, who is involved in the doing, and who owns the data (or to whom the data is relevant). Training of staff will be a crucial element of ensuring that data is secure, privacy obligations are not breached, and liability as an issue does not arise.

Significantly, even when no law is breached, your collection, collation, and use of big data will require planning and conscious thought. You should consider whether your proposed collection, collation, or use is permitted or constrained by the terms of the agreement regulating your membership of the group or the agreement that grants you access to the data. A breach of contractual obligations, even if innocent and unintended, may have very expensive unintended consequences.

As Jain et al. (2016) observed, the widespread use of big data is often in conflict with the needs for its privacy and security. Ensuring you engage with the specific legal obligations of your country, or those in which you or your business operates, will be vital as the first step in ensuring your legal compliance. The second will be remembering to read the contractual terms of access, use, or membership to make sure there are no contractual constraints or penalties for what you propose to do.

Notes

1 In Europe, data protection is a separate human right granted by Article 8 of the Charter of Fundamental Rights. See McDermott, Y. 2017. Conceptualizing the right to data protection in era of Big Data. *Big Data & Society*, January–June, 1–7.

2 "This new framework protects the fundamental rights of anyone in the EU whose personal data is transferred to the United States as well as bringing legal clarity for businesses relying on transatlantic data transfers." (European Commission, 2016).

3 For those interested in further reading in this area see, Perry, M. (ed.). 2016. *Global Governance of Intellectual Property in the 21st Century: Reflecting Policy through Change* (Springer International Publishing, Switzerland).

4 Any data accessed and/or collected and/or analyzed for the purpose of this book complied with all legal and contractual requirements.

References

Almeida, V., Doneda, D., & Monteiro, M. (2015). Governance challenges for the internet of things. *IEEE Internet Computing, 10*(2), 56–59.

Autorité de la concurrence and Bundeskartellant. (2016) Competition Law and Data, Report, May 10, 2016. www.autoritedelaconcurrence.fr/doc/reportcompetitionlawanddatafinal.pdf

Berne Convention for the Protection of Literary and Artistic Works (1886).

Charter of Fundamental Rights of the European Union (2012/C 326/02) of 26 October 2010.

Couldry, N. & Powell, A. (2014). Big data from the bottom up. *Big, Data & Society*, July–December, 1–5.

Cradduck, L. M. & McCullagh, A. J. (2007). Identifying the identity thief: Is it time for a (smart) Australia Card? *International Journal of Law and Information Technology, 16*(2), 1–34.

Data Protection Act 1998 (UK).

Directive 95/46/EC of the European Parliament and of the Council of 24 October 1995 on the protection of individuals with regard to the processing of personal data and on the free movement of such data.

Dudley-Nicholson, J. & Bickers, C. (2016). *Australia's 2016 Census had "significant and obvious oversights", report finds.* November 25, 2016. news.com.au. www.news.com.au/technology/online/australias-2016-census-had-significant-and-obvious-oversights-report-finds/news-story/6edcf8f897b2361965bd72683ee6edbe

European Commission. (2016, July 12). European Commission launches EU-U.S: Privacy Shield: A Stronger Protection for Transatalantic Data Flow. *Press Release*, Brussels. http://europa.eu/rapid/press-release_IP-16-2461_en.htm

Franceschi-Bicchierai, L. (2016, October 20). How Hackers Broke Into John Podesta and Colin Powell's Gmail Accounts. *Motherboard*. https://motherboard.vice.com/en_us/article/how-hackers-broke-into-john-podesta-and-colin-powells-gmail-accounts

Gregory, M. & Glance, D. (2013). *Security and the Networked Society.* New York: Springer.

Ishii, K. & Komukai, T. (2016). A Comparative Legal Study on Data Breaches in Japan, the U.S., and the U.K. *Technology and Intimacy: Choice or Coercion 12th IFIP TC 9 International Conference on Human Choice and Computers*, HCC12 2016, Salford, September 7–9, 2016.

Jain, P., Gyanchandani, M., & Khare, N. (2016). Big data privacy: A technological perspective and review. *Journal of Big Data, 3*, 25.

Kobus, T. et al. (2015). *2015 International Compendium of Data Privacy Laws*, Baker Hostetler.

Lodder, A. (2013). Ten commandments of internet law revisited: Basic principles for internet lawyers. *Information and Communications Technology Law, 22*(3), 264–276.

Marrakesh Agreement Establishing the World Trade Organization, Annex 1C, Agreement on Trade Related Aspects of Intellectual Property Rights, 15 April 1994, 33 I.L.M. (1994) 81 (TRIPS).

McDermott, Y. (2017). Conceptualising the right to data protection in era of big data. *Big Data & Society*, January–June, *4*(1), 1–7.

Morabito, V. (2015). *Big Data and Analytics.* Switzerland: Springer International Publishing.

OECD. (2013). The OECD Privacy Framework. www.oecd.org/sti/ieconomy/oecd_privacy_framework.pdf

Perry, M. (ed.). (2016). *Global Governance of Intellectual Property in the 21st Century: Reflecting Policy Through Change.* Switzerland: Springer International Publishing.

Ponemon Institute. (2011). *The security of cloud infrastructure: Survey of U.S. IT and compliance practitioners, Ponemon Institute LLC.* Research Report, November 2011.

Privacy Act 1974, 5 U.S.C. §552a.

Privacy Act 1988 (Aust).

Rahmani, A., Amine, A., & Hamou, R. (2016). Combination access control and de-identification for privacy preserving in big data. *International Journal of Information Security and Privacy, 10*(1), 1–27.

Raja, A. (2016). Online privacy and security concerns of consumers. *Information & Computer Security, 24*(4), 348–371.

Shah, M., Ahmed, J., & Soomro, J. (2016). Investigating the Identity Theft Prevention Strategies in M-Commerce. *International Conferences ITS, ICEdu Tech and STE 2016*, Melbourne, Australia, 59–66.

Smith, A. (2016). Little appetite for big data? *Journal of Property Investment & Finance, 34*(4). doi: 10.1108/jpif-04-2016-0024.

Sodiya, A. & Adegbuyi, B. (2016). A framework for protecting users' privacy in cloud. *International Journal of Information Security and Privacy, 10*(4), 33–43.

Svantesson, D. (2013). *Extraterritoriality in Data Privacy Law.* Denmark: Ex Tuto Publishing.

U.S. Department of Commerce. (2016). EU-U.S. Privacy Shield. www.privacyshield.gov/EU-US-Framework

World Intellectual Property Organization Copyright Treaty, adopted 20 December 1996, WIPO Doc. CRNR/DC/94.

14 The future of big data

In many regards the future of big data in real estate is much greater than its past; the changes, challenges, and opportunities that big data present are growing as quickly as the data itself. The future promises to be one where researchers have access to much more data, in many different forms, and the challenge is going to be discerning how to use it properly to better understand human behavior. Even as this volume goes to press, big data is affecting the way we market, evaluate, and invest in real estate. The tools used to explore it are becoming much more specialized, and the demand for the knowledge big data promises continues to grow.

As we reflect on the contribution big data has and is making, we would be remiss not to look forward and speculate on the future impact it may have on real estate markets, researchers, and analysts. While we acknowledge that predictions are often fraught with inaccuracies, oversights, and missteps, we willingly move forward to hopefully provide some general direction to the discipline. Some of our predictions are directed at the industry while others are focused on the research often carried out by university faculty and students. Either way the impact of big data has the potential to be extraordinary.

With that in mind we humbly prognosticate that:

1 Data is the new raw material (or as it is described in the *Economist*, "the oil of the digital era" [*Economist*, 2017]). This is more of an observation than a prediction and is something that researchers always knew (even if they did not describe it as such) since data is a fundamental component of quality research. What is different in the era of big data is a much broader recognition that the data that are being collected by firms, governments, and others can be used to increase efficiencies, broaden market appeal, and improve bottom lines. Even if these entities do not know how to use big data or what benefits it will provide, the perception that these data are valuable will partially drive the demand for better access, tools, and expertise to explore its potential.
2 The volume of data will continue to grow. At the risk of stating the obvious, we see no upward limit to the amount of data that will be created and we would be foolish to even contemplate it. What we can say with relative certainty is that the continual flood of data will force innovation at all levels.
3 As research becomes more acclimated to these new sources and types of data, we expect to see some retraction from the focus on the "big" part of big data. Data will be curated to such an extent that only the most meaningful components are extracted, reaffirming that quality data supersedes large quantities in almost every situation.

4 The tools used to decode big data will diversify as research needs diversify. Statistical software providers will be forced to frequently modify packages to include more user-friendly interfaces as data availability in a variety of formats expands. Continued competition from open source and freely available software will pressure for-profit providers to rethink how and what they offer clients. Without significant innovations, legacy contracts with for-profit software providers will give way to more open source providers who differentiate their products by supporting a broad user community.

5 The market for real estate data will continue to evolve. Real estate data are very specialized and efficient markets for these data are in their infancy. The number of firms that standardize and manage real estate data effectively is limited to just a few – it spans the border between a duopoly and an oligopoly. CoreLogic and Black Knight perform quite well in the U.S., but more competitors are needed to increase market efficiency and improve availability. We acknowledge there are very high barriers to entry (e.g., collecting data in non-disclosure states; data collection through field observation; data entry) but also see room for consolidators with a highly specialized focus that concentrate on less common yet equally informative types of data. Additionally, we see potential for new firms and products in the data curation space as the importance of quality data increases.

6 The specialization of real estate data will be maneuvered into more complex, collateral, and market leading (vis-a-vis market lagging) data to further a firm's competitive advantage. Blackrock, for example, is pioneering the innovative use of unstructured data by using aerial photography to better estimate changes in national economies, thereby getting forecasts of economic growth figures a full month before official statistics and beating the market. Accordingly, primary emphasis will be placed on measuring investor motivations on a real-time or nearly real-time basis to achieve similar market leading forecasts in the REIT, residential, and commercial real estate markets.

7 The demand for data scientists will increase much more quickly than the supply. Data science is one of the fastest growing industries in the world with a recent headline in Forbes stating "IBM Predicts Demand For Data Scientists Will Soar 28% By 2020." Forecasts estimate that 59% of the demand will come from finance and insurance, professional services, and information technology (Columbus, 2017). The resulting multi-level expertise required of data scientists will not only include statistics and computer analytics, it will also necessitate discipline-specific knowledge of real estate, finance, or whatever other area the researcher is investigating.

8 Universities that award degrees in real estate will be forced to refine or redevelop course offerings to include a strong data-driven research component. This may be as simple as the addition of a class or two or the introduction of a major/minor in information technology and real estate. If they wish to position themselves competitively, interested students should be advised to consider replacing the traditional combinations of real estate and finance with an academic plan that includes subjects in statistics, computer science, and real estate. Dual degree options could also be made available, which will greatly enhance the marketability of qualified students.

9 The privacy, ethical, and legal issues associated with big data will be an ongoing concern. There is no universal standard for information dissemination and what is known and what should be known about individuals is debatable. Even though

much real estate data is collected locally, distribution is an international business. Whether it is big data or small, conforming to the laws and customs of different countries requires a level of expertise many firms are unlikely to possess.

While there is little to hold back the rising tide of data, we are hopeful that this book provides the reader with enough insight that he or she can confidently and effectively engage in research that draws from these new and novel sources. The abundance of data is helping drive the innovation that produces exciting yet sound research, and the future applications are well beyond what we can envision here. It is now up to the researcher to take the direction we provide and make use of it in a way that benefits the academic, practitioner, investor, and user of real estate.

References

Columbus, L. (2017). IBM Predicts Demand for Data Scientists Will Soar 28% By 2020. *Forbes*, 13 May. www.forbes.com/sites/louiscolumbus/2017/05/13/ibm-predicts-demand-for-data-scientists-will-soar-28-by-2020/#6bb8dd3d7e3b, accessed on May 14, 2017.

Economist. (2017). The World's Most Valuable Resource Is No Longer Oil, But Data, 6 May. www.economist.com/news/leaders/21721656-data-economy-demands-new-approach-antitrust-rules-worlds-most-valuable-resource, accessed on May 18, 2017.

Index